Gathering Moss

Gathering Moss

A Natural and Cultural History of Mosses

by

Robin Wall Kimmerer

Oregon State University Press ⁂ Corvallis

For my family

Library of Congress Cataloging-in-Publication Data
Kimmerer, Robin Wall.
Gathering moss : a natural and cultural history of mosses /
by Robin Wall Kimmerer.
p. cm.
Includes bibliographical references (p.).
ISBN 0-87071-499-6 (alk. paper)
1. Mosses. 2. Mosses—Ecology. 3. Kimmerer, Robin Wall. I. Title.
QK537 .K56 2003
588'.2—dc21

2002151221

∞ This paper meets the requirements of ANSI/NISO Z39.48-1992
(Permanence of Paper).

 First edition 2003
Twenty-first printing 2023
Printed in the United States of America

Oregon State University
OSU Press

121 The Valley Library Corvallis OR 97331
541-737-3166 • fax 541-737-3170
www.osupress.oregonstate.edu

Table of Contents

Preface: *Seeing the World Through Moss-colored Glasses*

My first conscious memory of "science" (or was it religion?) comes from my kindergarten class, which met in the old Grange Hall. We all ran to press our noses to the frosty windows when the first intoxicating flakes of snow began to fall. Miss Hopkins was too wise a teacher to try and hold back the excitement of five-year-olds on the occasion of the first snow, and out we went. In boots and mittens, we gathered around her in the soft swirl of white. From the deep pocket of her coat she took a magnifying glass. I'll never forget my first look at snowflakes through that lens, spangling the wooly sleeve of her navy blue coat like stars in a midnight sky. Magnified tenfold, the complexity and detail of a single snowflake took me completely by surprise. How could something as small and ordinary as snow be so perfectly beautiful? I couldn't stop looking. Even now, I remember the sense of possibility, of mystery that accompanied that first glimpse. For the first time, but not the last, I had the sense that there was more to the world than immediately met the eye. I looked out at the snow falling softly on the branches and rooftops with a new understanding, that every drift was made up of a universe of starry crystals. I was dazzled by what seemed a secret knowledge of snow.The lens and the snowflake, were an awakening, the beginning of seeing. It's the time when I first had an inkling that the already gorgeous world becomes even more beautiful the closer you look.

Learning to see mosses mingles with my first memory of a snowflake. Just at the limits of ordinary perception lies another level in the hierarchy of beauty, of leaves as tiny and perfectly ordered as a snowflake, of unseen lives complex and beautiful. All it takes is attention and knowing how to look. I've found mosses to be a vehicle for intimacy with the landscape, like a secret knowledge of the forest. This book is an invitation into that landscape.

Three decades after my first look at mosses, I almost always have my hand lens around my neck. Its cord tangles with the leather thong of my medicine bag, in metaphor and in reality. The knowledge I have of plants has come from many sources, from the plants themselves, from my training as a scientist, and from an intuitive affinity for the traditional knowledge of my Potawatomi heritage. Long before I went to university to learn their scientific names, I regarded plants as my teachers. In college, the two perspectives on the life of plants, subject and object, spirit and matter, tangled like the two cords around my neck. The way I was taught plant science pushed my traditional knowledge of plants to the margins. Writing this book has been a process of reclaiming that understanding, of giving it its rightful place.

Our stories from the oldest days tell about the time when all beings shared a common language—thrushes, trees, mosses, and humans. But that language has been long forgotten. So we learn each other's stories by looking, by watching each other's way of living. I want to tell the mosses' story, since their voices are little heard and we have much to learn from them. They have messages of consequence that need to be heard, the perspectives of species other than our own. The scientist within me wants to know about the life of mosses and science offers one powerful way to tell their story. But it's not enough. The story is also about relationship. We've spent a long time knowing each other, mosses and I. In telling their story, I've come to see the world through moss-colored glasses.

In indigenous ways of knowing, we say that a thing cannot be understood until it is known by all four aspects of our being: mind, body, emotion, and spirit. The scientific way of knowing relies only on empirical information from the world, gathered by body and interpreted by mind. In order to tell the mosses' story I need both approaches, objective and subjective. These essays intentionally give voice to both ways of knowing, letting matter and spirit walk companionably side by side. And sometimes even dance.

Acknowledgements

I am grateful for the many good people who have shared in the creation of this book. Thanks to my father, Robert Wall, for the time spent looking at mosses and the craft that led to the beautiful drawings. It has been a joy to work together. I gratefully acknowledge also the permission to use the illustrations by the late Howard Crum, a great bryologist whose books and drawings have introduced mosses to so many. I thank Pat Muir and Bruce McCune for their hospitality and encouragement, Chris Anderson and Dawn Anzinger for their reading, the National Science Foundation and Oregon State University for their support during the sabbatical which made this book possible. The advice and support of Mary Elizabeth Braun and Jo Alexander at the Oregon State University Press have been much appreciated. The suggestions of reviewers Janice Glime, Kareen Sturgeon, members of my Bryophyte Ecology class at SUNY College of Environmental Science and Forestry, and many friends who have offered comment and support have been very helpful. Above all, I have been blessed with a loving family who create a habitat where good things can grow. I thank my mother for listening to my writing from the very beginning and creating a place for beauty, my father for bringing me into the woods and fields, my sisters and brother for their encouragement. I thank Jeff for believing in me at every step. I am especially thankful for the loving support and generosity of my daughters, Linden and Larkin, who have been my inspiration.

The Drawings

The figures on the following pages are used by permission from *Mosses of the Great Lakes Forest* by Howard Crum: 24, 38, 40, 41, 98, 122, 123, 128, 142

All other figures are by R. L. Wall.

The Standing Stones

Barefoot, I've walked this path by night for nearly twenty years, most of my life it seems, the earth pressing up against the arch of my foot. More often than not, I leave my flashlight behind, to let the path carry me home through the Adirondack darkness. My feet touching the ground are like fingers on the piano, playing from memory an old sweet song, of pine needles and sand. Without thinking, I know to step down carefully over the big root, the one by the sugar maple, where the garter snakes bask every morning. I whacked my toe there once, so I remember. At the bottom of the hill, where the rain washes out the path, I detour into the ferns for a few steps, avoiding the sharp gravel. The path rises over a ridge of smooth granite, where I can feel the day's warmth still lingering in the rock. The rest is easy, sand and grass, past the place where my daughter Larkin stepped into a nest of yellow jackets when she was six, past the thicket of striped maple where we once found a whole family of baby screech owls, lined up on a branch, sound asleep. I turn off toward my cabin, just at the spot where I can hear the spring dripping, smell its dampness, and feel its moisture rising up between my toes.

I first came here as a student, to fulfill my undergraduate requirement for field biology at the Cranberry Lake Biological Station. This is where I was first introduced to mosses, following Dr. Ketchledge through the woods, discovering mosses with the aid of the standard-issue hand lens, the Wards Scientific Student Model borrowed from the stock room, that I wore around my neck on a dirty string. I knew I was committed when, at the end of the class, I spent a part of my sparse college savings to send off for a professional-grade Bausch and Lomb lens like his.

I still have that lens, and wear it on a red cord, as I take my own students along the trails at Cranberry Lake, where I returned to join the faculty and to eventually become the Bio Station Director. In all those years, the mosses haven't changed nearly as much as I have. That patch of *Pogonatum* that Ketch showed us along the Tower Trail is still there. Each summer I stop for a closer look and wonder at its longevity.

These past few summers, I've been conducting research on rocks, trying to learn what I can about how communities form, by observing the way that moss species gather together on boulders. Each boulder stands apart like a desert island in a swelling sea of forest. Its only inhabitants are mosses. We're trying to figure out why on one rock ten or more species of moss may comfortably coexist, while a nearby boulder, outwardly the same, is completely dominated by just a single moss, living alone. What are the conditions that foster diverse communities rather than isolated individuals? The question is very complex for mosses, let alone for humans. By summer's end, we should have a tidy little publication, our scholarly contribution to the truth about rocks and mosses.

Glacial boulders are scattered over the Adirondacks, the round tumbled granite left behind by the ice ten thousand years ago. Their mossy bulk makes the forests seem primeval and yet I know how much the scenery has changed around them, from the day they were stranded in a barren plain of glacial outwash, to the thick maple woods that surround them today.

Most of the boulders reach only to my shoulders, but for some we need a ladder to survey them completely. My students and I wrap measuring tape around their girth. We record light and pH, collect data on the number of crevices and the depth of the thin skin of humus. We carefully catalog the positions of all the moss species, calling out their names. *Dicranum scoparium. Plagiothecium denticulatum.* The student struggling to record all this begs for shorter names. But mosses don't usually have common names, for no one has bothered with them. They have only scientific names, conferred with legalistic formality according to protocol set up by Carolus Linnaeus, the great plant taxonomist. Even his own name, Carl Linne, the name his Swedish mother had given him, was Latinized in the interest of science.

A good number of the rocks around here have names, and people use them for reference points around the lake: Chair Rock, Gull Rock, Burnt Rock, Elephant Rock, Sliding Rock. Each name calls up a story, and connects us to the past and present of this place every time we say it. My daughters, being raised in a place where they simply assume that all rocks have names, christen their own: Bread Rock, Cheese Rock, Whale Rock, Reading Rock, Diving Rock.

The names we use for rocks and other beings depends on our perspective, whether we are speaking from the inside or the outside of the circle. The name on our lips reveals the knowledge we have of each other, hence the sweet secret names we have for the ones we love. The names we give ourselves are a powerful form of self-determination, of declaring ourselves sovereign territory. Outside the circle, scientific names for mosses may suffice, but within the circle, what do they call themselves?

One of the charms of the Bio Station is that it doesn't change much from summer to summer. We can put it on each June, like a faded flannel shirt still smelling of last summer's woodsmoke. It's the bedrock of our lives, our true home, a constant amidst so many other changes. There hasn't been a summer when the parulas didn't nest in the spruces by the dining hall. In mid-July, before the blueberries ripen, a bear will regularly wander through camp, hungry. Beavers swim like clockwork past the front dock, twenty minutes after sunset, and the morning mist always hangs longest in the southern coll on Bear Mountain. Oh, sometimes things change. In a hard winter, the ice might shift the driftwood on the shore. Once a silvery old log, the one with a branch like a heron's neck, got moved sixty feet down the bay. And one summer, the sapsuckers ended up nesting in a different tree, after the top blew out of the rotten old aspen in a gale. Even the changes make familiar patterns, like the marks of waves on the sand, the way the lake can go from flat calm to three-foot rollers, the way the aspen leaves sound hours before a rain, the way the texture of the evening clouds foretells the next day's winds. I find strength and comfort in this physical intimacy with the land, a sense of knowing the names of the rocks and knowing my place in the world. On this wild shore, my internal landscape is a near-perfect reflection of the external world.

So, I was stunned by what I saw today, on what had always been a familiar trail, a few miles down the shore from my cabin. It stopped me in my tracks. Disoriented, I caught my breath, glancing all around to reassure myself that I was still on the same trail and hadn't wandered into some twilight zone, where things are not as they seem. I've walked this path more times than I can tell, and yet it was only today that I was able to see them: five stones, each the size of a school bus, lying together in a pile, their curves fitting together like an old married couple secure in each other's arms. The glacier must have pushed them into this loving conformation and then moved on. I circle all around the pile, in silence, brushing my fingertips over its mosses.

On the eastern side, there is an opening, a cave-like darkness between the rocks. Somehow, I knew it would be there. This door which I have never seen before looks strangely familiar. My family comes from the Bear Clan of the Potawatomi. Bear is the holder of medicine knowledge for the people and has a special relationship with plants. He is the one who calls them by name, who knows their stories. We seek him for a vision, to find the task we were meant for. I think I'm following a Bear.

The landscape itself seems alert, with every detail in unnaturally sharp focus. I stand in an island of surreal quiet where time feels as weighty as the rocks. And yet, when I shake my head, clearing my vision, I can hear the familiar whoosh of waves on the shore and the redstarts chittering over my head. The cave draws me inward, on my hands and knees into the dark, beneath the tons of rock, imagining the den of a bear. I creep ahead, the rock rough against my bare arms. Around a turn the light from outside disappears behind me. I breathe in the coolness, and there is no scent of bear, just the soft ground and smell of granite. Feeling with my fingers, I go forward, but I don't quite know why. The cave floor slants downward, dry and sandy as if the rains never penetrate this far. Ahead of me, around another corner, the tunnel rises. There is green forest light ahead, so I push on. I think I must have crawled through a passage leading beneath this pile of rock and out the other side. I wriggle from the tunnel and find myself not in the woods at all. Instead, I emerge into a tiny grass-filled meadow, a circle enclosed by the walls of the stones. It is a room, a light-filled room like a round eye looking up into the blueness of the sky. Indian paintbrush is in bloom and hay-scented fern borders the ring of the standing stones. I am inside the circle. There are no openings save the way that I have come and I sense that entrance closing behind me. I look all around the ring but I can no longer see the opening in the rock. At first I'm afraid, but the grass smells warm in the sunshine and the walls drip with mosses. How odd to still hear the redstarts calling in the trees outside, in a parallel universe that dissipates like a mirage as the mossy walls enclose me.

Within the circle of stones, I find myself unaccountably beyond thinking, beyond feeling. The rocks are full of intention, a deep presence attracting life. This is a place of power, vibrating with energy exchanged at a very long wavelength. Held in the gaze of the rocks, my presence is acknowledged.

The rocks are beyond slow, beyond strong, and yet yielding to a soft green breath as powerful as a glacier, the mosses wearing away their surfaces, grain by grain bringing them slowly back to sand. There is an ancient conversation going on between mosses and rocks, poetry to be sure. About light and shadow and the drift of continents. This is what has been called the "dialectic of moss on stone—an interface of immensity and minuteness, of past and present, softness and hardness, stillness and vibrancy, yin and yang."[1] The material and the spiritual live together here.

Moss communities may be a mystery to scientists, but they are known to one another. Intimate partners, the mosses know the contours of the rocks. They remember the route of rainwater down a crevice, the way I remember the path to my cabin. Standing inside the circle, I know that mosses have their own names, which were theirs long before Linnaeus, the Latinized namer of plants. Time passes.

I don't know how long I was gone, minutes or hours. For that interval, I had no sensation of my own existence. There was only rock and moss. Moss and rock. Like a hand laid gently on my shoulder, I come back to myself and look around. The trance is broken. I can hear the redstarts again, calling overhead. The encircling walls are radiant with mosses of every kind, and I see them again, as if for the first time. The green and the gray, the old and the new in this place and in this time, they rest together for this moment between glaciers. My ancestors knew that rocks hold the Earth's stories, and for a moment I could hear them.

My thoughts feel noisy here, an annoying buzz disrupting the slow conversation among the stones. The door in the wall has reappeared

---

1. Schenk, H. *Moss Gardening*, 1999.

and time starts to move again. An opening into this circle of stones was made, and a gift given. I can see things differently, from the inside of the circle as well as from the outside. A gift comes with responsibility. I had no will at all to name the mosses in this place, to assign their Linnean epithets. I think the task given to me is to carry out the message that mosses have their own names. Their way of being in the world cannot be told by data alone. They remind me to remember that there are mysteries for which a measuring tape has no meaning, questions and answers that have no place in the truth about rocks and mosses.

The tunnel seems easier on the way out. This time I know where I am going. I look back over my shoulder at the stones and then set my feet to the familiar path for home. I know I'm following a Bear.

Learning to See

After four hours at 32,000 feet, I've finally succumbed to the stupor of a transcontinental flight. Between takeoff and landing, we are each in suspended animation, a pause between chapters of our lives. When we stare out the window into the sun's glare, the landscape is only a flat projection with mountain ranges reduced to wrinkles in the continental skin. Oblivious to our passage overhead, other stories are unfolding beneath us. Blackberries ripen in the August sun; a woman packs a suitcase and hesitates at her doorway; a letter is opened and the most surprising photograph slides from between the pages. But we are moving too fast and we are too far away; all the stories escape us, except our own. When I turn away from the window, the stories recede into the two-dimensional map of green and brown below. Like a trout disappearing into the shadow of an overhanging bank, leaving you staring at the flat surface of the water and wondering if you saw it at all.

I put on my newly acquired and still frustrating reading glasses and lament my middle-aged vision. The words on the page float in and out of focus. How is it possible that I can no longer see what was once so plain? My fruitless strain to see what I know is right in front of me reminds me of my first trip into the Amazon rain forest. Our indigenous guides would patiently point out the iguana resting on a branch or the toucan looking down at us through the leaves. What was so obvious to their practiced eyes was nearly invisible to us. Without practice, we simply couldn't interpret the pattern of light and shadow as "iguana" and so it remained right before our eyes, frustratingly unseen.

We poor myopic humans, with neither the raptor's gift of long-distance acuity, nor the talents of a housefly for panoramic vision. However, with our big brains, we are at least aware of the limits of our vision. With a degree of humility rare in our species, we acknowledge there is much that we can't see, and so contrive remarkable ways to observe the world. Infrared satellite imagery, optical telescopes, and the Hubbell space telescope bring vastness within our visual sphere. Electron microscopes let us wander the remote universe of our own cells. But at the middle scale, that of the unaided eye, our senses seem to be strangely dulled. With sophisticated technology, we strive to see what is beyond us, but are often blind to the myriad sparkling facets that lie so close at hand. We think we're *seeing* when we've only scratched the surface. Our acuity at this middle scale seems diminished, not by any failing of the eyes, but by the willingness of the mind. Has the power of our devices led us to distrust our unaided eyes? Or have we become dismissive of what takes no technology but only time and patience to perceive? Attentiveness alone can rival the most powerful magnifying lens.

I remember my first encounter with the North Pacific, at Rialto Beach on the Olympic Peninsula. As a landlocked botanist, I was anticipating my first glimpse of the ocean, craning my neck around every bend in the winding dirt road. We arrived in a dense gray fog that clung to the trees and beaded my hair with moisture. Had the skies been clear we would have seen only what we expected: rocky coast, lush forest, and the broad expanse of the sea. That day, the air was opaque and the backdrop of coastal hills was visible only when the spires of Sitka Spruce briefly emerged from the clouds. We knew the ocean's presence only by the deep roar of the surf, out beyond the tidepools. Strange, that at the edge of this immensity, the world had become very small, the fog obscuring all but the middle distance. All my pent-up desire to see the panorama of the coast became focussed on the only things that I could see, the beach and the surrounding tidepools.

Wandering in the grayness, we quickly lost sight of each other, my friends disappearing like ghosts in just a few steps. Our muffled voices knit us together, calling out the discovery of a perfect pebble, or the intact shell of a razor clam. I knew from poring over field guides in anticipation of the trip that we "should" see starfish in the tidepools, and this would be my first. The only starfish I'd ever seen was a dried one in a zoology class and I was eager to see them at home where they belonged. As I looked among the mussels and the limpets, I saw none. The tidepools were encrusted with barnacles and exotic-looking algae, anemones, and chitons enough to satisfy the curiosity of a novice tidepooler. But no starfish. Picking my way over the rocks, I pocketed fragments of mussel shells the color of the moon, and tiny sculpted pieces of driftwood, looking continuously. No starfish. Disappointed, I straightened up from the pools to relieve the growing stiffness in my back, and suddenly—I saw one. Bright orange and clinging to a rock right before my eyes. And then it was as if a curtain had been pulled away and I saw them everywhere. Like stars revealing themselves one by one in a darkening summer night. Orange stars in the crevices of a black rock, speckled burgundy stars with outstretched arms, purple stars nestled together like a family huddled against the cold. In a cascade of discovery, the invisible was suddenly made visible.

A Cheyenne elder of my acquaintance once told me that the best way to find something is not to go looking for it. This is a hard concept for a scientist. But he said to watch out of the corner of your eye, open to possibility, and what you seek will be revealed. The revelation of suddenly seeing what I was blind to only moments before is a sublime experience for me. I can revisit those moments and still feel the surge of expansion. The boundaries between my world and the world of another being get pushed back with sudden clarity, an experience both humbling and joyful.

The sensation of sudden visual awareness is produced in part by the formation of a "search image" in the brain. In a complex visual landscape, the brain initially registers all the incoming data, without critical evaluation. Five orange arms in a starlike pattern, smooth black rock, light and shadow. All this is input, but the brain does not immediately interpret the data and convey their meaning to the conscious mind. Not until the pattern is repeated, with feedback from the conscious mind, do we know what we are seeing. It is in this way that animals become skilled detectors of their prey, by differentiating complex visual patterns into the particular configuration that means food. For example, some warblers are very successful predators when a certain caterpillar is at epidemic numbers, sufficiently abundant to produce a search image in the bird's brain. However, the very same insects may go undetected when their numbers are low. The neural pathways have to be trained by experience to process what is being seen. The synapses fire and the stars come out. The unseen is suddenly plain.

At the scale of a moss, walking through the woods as a six-foot human is a lot like flying over the continent at 32,000 feet. So far above the ground, and on our way to somewhere else, we run the risk of missing an entire realm which lies at our feet. Every day we pass over them without seeing. Mosses and other small beings issue an invitation to dwell for a time right at the limits of ordinary perception. All it requires of us is attentiveness. Look in a certain way and a whole new world can be revealed.

My former husband used to teasingly deride my passion for mosses, saying that mosses were just decoration. To him, mosses were merely the wallpaper of the forest, providing ambience for his photographs of trees. A carpet of mosses does in fact provide this lustrous green light. But, focus the lens on the mossy wallpaper itself and the green blur of the background resolves itself into sharp focus and an entirely new dimension appears. That wallpaper, which seemed at first glance to be of uniform weave, is in fact a complex tapestry, a brocaded surface of intricate pattern. The "moss" is many different mosses, of widely divergent forms. There are fronds like miniature ferns, wefts like ostrich plumes, and shining tufts like the silky hair of a baby. A close encounter with a mossy log always makes me think of entering a fantasy fabric shop. Its windows overflow with rich textures and colors that invite you closer to inspect the bolts of cloth arrayed before you. You can run your fingertips over a silky drape of *Plagiothecium* and finger the glossy *Brotherella* brocade. There are dark wooly tufts of *Dicranum*, sheets of golden *Brachythecium*, and shining ribbons of *Mnium*. The yardage of nubbly brown *Callicladium* tweed is shot through with gilt threads of *Campylium*. To pass hurriedly by without looking is like walking by the Mona Lisa chatting on a cell phone, oblivious.

Draw closer to this carpet of green light and shadow, and slender branches form a leafy arbor over sturdy trunks, rain drips through the canopy, and scarlet mites roam over the leaves. The architecture of the surrounding forest is repeated in the form of the moss carpet, the fir forest and the moss forest mirroring each other. Let your focus shift to the scale of a dewdrop, the forest landscape now becomes the blurred wallpaper, only a backdrop to the distinctive moss microcosm.

Learning to see mosses is more like listening than looking. A cursory glance will not do it. Straining to hear a faraway voice or catch a nuance in the quiet subtext of a conversation requires attentiveness, a filtering of all the noise, to catch the music. Mosses are not elevator music; they are the intertwined threads of a Beethoven quartet. You can look at mosses the way you can listen deeply to water running over rocks. The soothing sound of a stream has many voices, the soothing green of mosses likewise. Freeman House writes of stream sounds; there is the rushing tumble of the stream running over itself, the splashing against rocks. Then, with care and quiet, the individual tones can be discerned in the fugue of stream sound. The slip of water over a boulder, octaves above the deep tone of shifting gravel, the gurgle of the channel sluicing between rocks, the bell-like notes of a drop falling into a pool. So it is with looking at mosses. Slowing down and coming close, we see patterns emerge and expand out of the tangled tapestry threads. The threads are simultaneously distinct from the whole, and part of the whole.

Knowing the fractal geometry of an individual snowflake makes the winter landscape even more of a marvel. Knowing the mosses enriches our knowing of the world. I sense the change as I watch my bryology students learn to see the forest in a whole new way.

I teach bryology in the summer, wandering through the woods, sharing mosses. The first few days of the class are an adventure as my students start to distinguish one moss from another, first by naked eye and then by hand lens. I feel like a midwife to an awakening, when they first discern that a mossy rock is covered not with "moss" but with twenty different kinds of moss, each one with its own story.

On the trail and in the lab, I like to listen to my students talk. Day by day, their vocabulary stretches and they proudly refer to leafy green shoots as "gametophytes" and the little brown thingamajigs on top of the moss are dutifully referred to as "sporophytes." The upright, tufted mosses become "acrocarps," the horizontal fronds are "pleurocarps." Having words for these forms makes the differences between them so much more obvious. With words at your disposal, you can see more clearly. Finding the words is another step in learning to see.

Another dimension and another lexicon open when the students start putting the mosses under the microscope. Individual leaves are removed by painstaking dissection and placed on a glass slide for detailed examination. Magnified twenty-fold, the surfaces of the leaves are beautifully sculpted. The light shining brightly through single cells illuminates their elegant shapes. Time can vanish in exploring these places, like wandering through an art gallery of unexpected forms and colors. Sometimes, I look up from my microscope at the end of an hour, and I'm taken aback at the plainness of the ordinary world, the drab and predictable shapes.

I find the language of microscopic description compelling in its clarity. The edge of a leaf is not simply uneven; there is a glossary of specific words for the appearance of a leaf margin: *dentate* for large, coarse teeth, *serrate* for a sawblade edge, *serrulate* if the teeth are fine and even, *ciliate* for a fringe along the edge. A leaf folded by accordion pleats is *plicate*, *complanate* when flattened as if squashed between two pages of a book. Every nuance of moss architecture has a word. The students exchange these words like the secret language of a fraternity, and I watch the bond between them grow. Having the words also creates an intimacy with the plant that speaks of careful observation. Even the surfaces of individual cells have their own descriptors—*mammillose* for a breast-like swelling, *papillose* for a little bump, and *pluripapillose* when there are enough bumps to look like chicken pox. While they may initially seem like arcane technical terms, these words have life to them. What better word for a thick, round shoot, swelling with water than *julaceous*?

Mosses are so little known by the general public that only a few have been given common names. Most are known solely by their scientific Latin names, a fact which discourages most people from attempting to identify them. But I like the scientific names, because they are as beautiful and intricate as the plants they name. Indulge yourself in the words, rhythmic and musical, rolling off your tongue: *Dolicathecia striatella, Thuidium delicatulum, Barbula fallax.*

Knowing mosses, however, does not require knowing their scientific names. The Latin words we give them are only arbitrary constructs. Often, when I encounter a new moss species and have yet to associate it with its official name, I give it a name which makes sense to me: green velvet, curly top, or red stem. The word is immaterial. What seems to me to be important is recognizing them, acknowledging their individuality. In indigenous ways of knowing, all beings are recognized as non-human persons, and all have their own names. It is a sign of respect to call a being by its name, and a sign of disrespect to ignore it. Words and names are the ways we humans build relationship, not only with each other, but also with plants.

The word "moss" is commonly applied to plants which are not actually mosses. Reindeer "moss" is a lichen, Spanish "moss" is a flowering plant, sea "moss" is an alga, and club "moss" is a lycophyte. So what *is* a moss? A true moss or bryophyte is the most primitive of land plants. Mosses are often described by what they lack, in comparison to the more familiar higher plants. They lack flowers, fruits, and seeds and have no roots. They have no vascular system, no xylem and phloem to conduct water internally. They are the most simple of plants, and in their simplicity, elegant. With just a few rudimentary components of stem and leaf, evolution has produced some 22,000 species of moss worldwide. Each one is a variation on a theme, a unique creation designed for success in tiny niches in virtually every ecosystem.

Looking at mosses adds a depth and intimacy to knowing the forest. Walking in the woods, and discerning the presence of a species from fifty paces away, just by its color, connects me strongly to the place. That certain green, the way it catches the light, gives away its identity, like recognizing the walk of a friend before you can see their face. Just as you can pick out the voice of a loved one in the tumult of a noisy room, or spot your child's smile in a sea of faces, intimate connection allows recognition in an all-too-often anonymous world. This sense of connection arises from a special kind of discrimination, a search image that comes from a long time spent looking and listening. Intimacy gives us a different way of seeing, when visual acuity is not enough.

The Advantages of Being Small: Life in the Boundary Layer

The wailing toddler attached to the end of my arm earns me a disapproving look from a sour-faced lady. My niece is inconsolable, because I made her hold my hand when we crossed the street. She is in full voice now, yelling, "I am not too little, I want to be big!" If she only knew how quickly her wish would come true. Back in the car, after she has whined through the ignominy of being buckled into her car seat, I try to have a reasonable talk with her, reminding her of the advantages of being small. She can fit in the secret fort under the lilac bush, and hide from her brother. What about stories in grandma's lap? But, she's not buying it. She falls asleep on the way home, clutching her new kite, a stubborn pout still on her face.

I brought a moss-covered rock to her pre-school for a science show and tell. I asked the kids at pre-school what a moss was. They skipped right over the question of animal, vegetable, or mineral and got directly to the most salient feature; mosses are small. Kids recognize that right away. This most obvious attribute has tremendous consequences for the way mosses inhabit the world.

Mosses are small because they lack any support system to hold them upright. Large mosses occur mostly in lakes and streams, where the water can support their weight. Trees stand tall and rigid because of their vascular tissue, the network of xylem, thick-walled tubular cells that conduct water within the plant like wooden plumbing. Mosses are the most primitive of plants and lack any such vascular tissue. Their slender stems couldn't support their weight if they were any taller. This same lack of xylem means that they can't conduct water from the soil to leaves at the top of the shoot. A plant more than a few centimeters high can't keep itself hydrated.

However, being small doesn't mean being unsuccessful. Mosses are successful by any biological measure—they inhabit nearly every ecosystem on earth and number as many as 22,000 species. Like my niece finding small places to hide, mosses can live in a great diversity of small microcommunities where being large would be a disadvantage. Between the cracks of the sidewalk, on the branches of an oak, on the back of a beetle, or on the ledge of a cliff, mosses can fill in the empty spaces left between the big plants. Beautifully adapted for life in miniature, mosses take full advantage of being small, and grow beyond their sphere at their peril.

With extensive root systems and shading canopies, trees are the undisputed dominants of the forest. Their competitive superiority and heavy leaf fall are no match for mosses. One consequence of being small is that competing for sunlight is simply not possible—the trees will always win. So mosses are usually limited to life in the shade, and they flourish there. The type of chlorophyll in their leaves differs from their sun-loving counterparts, and is fine-tuned to absorb the wavelengths of light that filter through the forest canopy.

Mosses are prolific under the moist shaded canopy of evergreens, often creating a dense carpet of green. But in deciduous forests, autumn makes the forest floor virtually uninhabitable by mosses, smothering them under a dark wet blanket of falling leaves. Mosses find a refuge from the drifting leaves on logs and stumps which rise above the forest floor like buttes above the plain. Mosses succeed by inhabiting places that trees cannot, hard, impermeable substrates such as rocks and cliff faces and bark of trees. But with elegant adaptation, mosses don't suffer from this restriction; rather, they are the undisputed masters of their chosen environment.

Mosses inhabit surfaces: the surfaces of rocks, the bark of trees, the surface of a log, that small space where earth and atmosphere first make contact. This meeting ground between air and land is known as the boundary layer. Lying cheek to cheek with rocks and logs, mosses are intimate with the contours and textures of their substrate. Far from being a liability, the size of mosses allows them to take advantage of the unique microenvironment created within the boundary layer.

What is this interface between atmosphere and earth? Every surface, be it as small as a leaf or as large as a hill, possesses a boundary layer. We've all experienced it in very simple ways. When you lie on the ground on a sunny summer afternoon to look up and watch the clouds go by, you place yourself in the boundary layer of the Earth's surface. When you are flat on the ground, the wind speed is reduced; you can scarcely feel the breeze that would ruffle your hair if you were standing up. It's warm down there as well; the sun-warmed ground radiates heat back at you, and the lack of breeze at the surface lets the heat linger. The climate right next to the ground is different from the one six feet above. The effect that we feel lying on the ground is repeated over every surface, large and small.

Air seems insubstantial, but it interacts in interesting ways with the things it touches, much as moving water interacts with the contours of the riverbed. As moving air passes over a surface like a rock, the surface changes the behavior of the air. Without obstacles, the air would tend to move smoothly in a linear path called *laminar flow*. If we could see it, it would look like water flowing freely in a smooth deep river. But as the air encounters a surface, friction tugs at the moving air and slows it down. You see this in the flow of water; when a river meets a rocky bottom or logs fallen in its path, the water slows. As the laminar flow is disrupted by the drag of the surface, the air stream becomes separated into layers of different speed. There is swiftly moving air aloft, flowing in a smooth sheet. Beneath it lies a zone of turbulence, where the air swirls and eddies as it encounters obstacles. Down toward the surface,

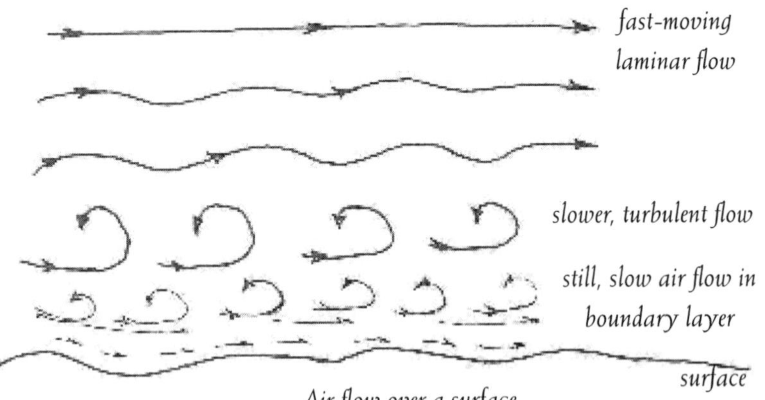

Air flow over a surface

the air becomes progressively slower and slower until, immediately adjacent to the surface, the air is perfectly still, captured by the friction with the surface itself. It is this layer of still air that you experience while lying on the ground.

At a larger scale, I encounter these layers of air every spring. On the first mild day in April, our beautiful kites that have been hanging draped with cobwebs on the porch all winter rustle in the breeze and remind us of blue sky. So, we take them out to play in the boundary layer. In our sheltered valley, the breezes are seldom strong enough to immediately catch the big dragon kites that the kids and I love. So we run crazily back and forth over the back pasture, dodging cow pies and trying to generate enough wind to carry the kite upward. Close to the surface of the earth, the winds are too slow to support the kites' weight. They are trapped, beyond the reach of the breeze. Only when our mad dashes loft one of the kites up to escape the layer of still air does it pull and dance on the string. Its wild pitches and threatened crashes show that it has ascended into the turbulent zone. And then at last, the kite's string pulls taut and the red and yellow dragon sails into the freely moving air above. Kites were made for the airy zone of laminar flow; mosses were made for the boundary layer.

Our pasture is littered with rocks left by the glacier, and I stop to sit on one and spool out the kite string, listening to meadowlarks. The rock is warm from the sun and softened by mosses. I can imagine the pattern of air, flowing smoothly around it until it encounters the surface, where the mosses live. The sun's warmth gets trapped in the tiny layer of

Pattern of air flow over a moss carpet

still air. Since the air is nearly motionless, it acts as an insulating layer, much like the dead space in a storm window, which forms a barrier to heat exchange. The spring breeze around me is chilly, but the air right at the surface of the rock is much warmer. Even on a day when the temperature is below freezing, the mosses on a sunlit rock may be bathed in liquid water. By being small, mosses can live in that boundary layer, like a floating greenhouse hovering just above the rock surface.

The boundary layer traps not only heat, but water vapor, as well. Moisture evaporating from the surface of a damp log is captured in the boundary layer, creating a humid zone in which the mosses flourish. Mosses can grow only when they are moist. As soon as they dry out, photosynthesis must cease, and growth is halted. The right conditions for growth can be infrequent, and so mosses grow very slowly. Living within the confines of the boundary layer prolongs the window of opportunity for growth, by keeping the wind from stealing the moisture. Being small enough to live within the boundary layer allows the mosses to experience a warm, moist habitat unknown by the larger plants.

The boundary layer can also hold gases other than water vapor. The chemical composition of the atmosphere in the slim boundary layer of a log differs considerably from that of the surrounding forest. The decaying log is inhabited by a myriad of microorganisms. Fungi and bacteria are constantly at work degrading the log, with an outcome as sure as that of a wrecking ball. The continual work of the decomposers slowly turns the solid log to crumbling humus and releases vapors rich in carbon dioxide, which is also trapped in the boundary layer. The ambient atmosphere has a carbon dioxide concentration of approximately 380 parts per million. But the boundary layer above a log may contain up to ten times that amount. Carbon dioxide is the raw material of photosynthesis, and is readily absorbed into the moist leaves of the mosses. Thus, the boundary layer can provide not only a favorable microclimate for moss growth, but also an enhanced supply of carbon dioxide, the raw material for photosynthesis. Why live anywhere else?

Being small enough to live in the boundary layer is a distinct advantage. Mosses have found the microhabitats where their size becomes an asset. The growth of a moss would be sharply curtailed if the shoots grew too tall and into the drying air of the turbulent zone. We might predict therefore that all mosses are uniformly small, corresponding to the limits imposed by the boundary layer. However, mosses exhibit a tremendous range in height, equivalent to the height disparity between a blueberry bush and a redwood. They range from tiny crusts only a millimeter high to lush wefts that can be up to ten centimeters tall. These differences in stature can usually be traced to differences in the depth of the boundary layer in the particular habitat. The boundary layer on a rock face exposed to wind and full sun is quite thin. Hence, the mosses of such arid places must be very small in order to stay within the protective boundary layer. In contrast, mosses on a rock in a moist forest can grow much taller and still remain within a favorable microclimate, because the boundary layer of the rock is under the umbrella of the boundary layer of the forest itself. The trees slow the wind and their shade reduces evaporation, buffering the area against the drying atmosphere. In a humid rainforest, the mosses can be lush and tall. The larger the boundary layer, the larger the moss can be.

Mosses can also control the depth of their own boundary layers by changes in their shape. Any feature of a surface that increases friction with moving air can slow the air and create a thicker boundary layer. A roughened surface slows the passage of air more effectively than a smooth one. Imagine being caught in a fierce prairie blizzard with strong winds blasting sheets of snow against your face. To escape the force of the winds, you lie down, taking refuge in the shelter of the earth's boundary layer. Given a choice, would you be warmer lying in the open or in a field of tall grass? The projection of the tall grass into the moving air stream slows the air and makes a larger boundary layer, helping to conserve your body heat. Mosses utilize this same principle to enlarge the boundary layers above them. The surface textures of a moss itself can create resistance to airflow. The greater the resistance, the deeper the boundary layer. Like a tall grassy field in miniature, moss shoots exhibit adaptations that slow air movement. Many moss species have long narrow leaves held upright to slow the airflow around them. Moreover, the leaves of mosses in dry sites often possess dense hairs, long reflective leaf tips, or minuscule spines. These extensions from the leaf surface also slow the moving air and reduce evaporation of essential moisture by creating a thicker boundary layer.

In arid zones, mosses often rely on dew for their daily ration of water. The interplay of the atmosphere and the rock surface creates the conditions for dew formation. At night, when the sun's heat dissipates, the temperature differential between the rock surface (which has retained some warmth) and the air may provide a site for condensation of water. A thin film of dew is created right at the air-rock interface, where it can be readily absorbed by the mosses. Only a very small being can take advantage of such a thin and evanescent supply of moisture in the desert, living on dew.

The safe and balmy realm of the boundary layer provides a secure refuge for mosses. But the very same nurturing environment that sustains growth to maturity poses a problem for the next generation. Like my niece, mosses eventually need to escape from protection by their elders and find their own places. Mosses reproduce by the formation of spores, tiny powdery propagules that require wind to carry them far afield. Most spores can't germinate in the leafy carpet of their own parents, so getting away is imperative. Air currents in the still air of the boundary layer are not sufficient to disperse them. So, to catch a breeze and help them leave the home territory, mosses elevate their spores on long setae, stalks that poke up above the boundary layer. The rapidly maturing sporophytes are thrust up through the boundary layer and into the turbulent zone like a kite on the wind. Here vortices of air swirl around the capsules, pulling out the spores and carrying them off to new habitats. Like the young of every species they escape the restrictions of their elders and seek out the freedom of the wide-open spaces.

The length of the seta or stalk is strongly correlated with the depth of the boundary layer. The seta of a forest moss must be quite tall to escape the boundary layer and catch the light breeze that moves over the forest floor. In contrast, mosses of open sites where the boundary layer is thin typically have short setae.

Mosses take possession of spaces from which other plants are excluded by their size. Their ways of being are a celebration of smallness. They succeed by matching the unique properties of their form to the physical laws of interaction between air and earth. In being small, their limitation is their strength. Try telling that to my niece.

Back to the Pond

I shiver in the damp breeze, but I can't bring myself to close the window on this April night that is sliding off the cusp of winter into spring. The faint sound of the peepers flows in with the cold air, but it's not enough. I need more. So I go downstairs and slip my down jacket over my nightgown, slide bare feet into my Sorels and leave the woodstove's warmth behind in the kitchen. With bootlaces dragging through patches of remnant snow, I tromp up to the pond above the farmhouse, breathing in the scents of wet ground. I'm pulled by the sound. Coming closer feels like walking into a crescendo, rising with the chorus of voices. I shiver again. The air literally throbs with the massed calls of peepers, vibrating the nylon shell of my jacket. I wonder at the power of these calls, bringing me from sleep and bringing the peepers back to the pond. Do we share some common language that draws us both to this place? The peepers have their own plan. What is it that brings me here to stand like a rock in this river of sound?

Their ringing calls summon all the local peepers to this gathering place, for mass fertilization in the rites of spring. Females will squeeze their eggs out into the shallow water, where males cover them with milky drifts of sperm. Surrounded by a gelatinous mass, the eggs will mature to tadpoles and become adult frogs by summer's end, long after their parents have hopped back to the woods. Spring peepers spend most of their adult lives as solitary tree frogs, travelling the forest floor. As far afield as they may venture, they must all return to water to reproduce. All amphibians are tethered to the pond by their evolutionary history, the most primitive vertebrates to make the transition from the aquatic life of their ancestors to life on land.

Mosses are the amphibians of the plant world. They are the evolutionary first step toward a terrestrial existence, a halfway point between algae and higher land plants. They have evolved some rudimentary adaptations to help them survive on land, and can survive even in deserts. But, like the peepers, they must return to water to breed. Without legs to carry them, mosses have to recreate the primordial ponds of their ancestors within their branches.

The next afternoon, I return to the now quiet pond, looking for some marsh marigolds to cook up for dinner. Bending to gather the leaves, I see the aftermath of last night, masses of eggs lying in the sunlit shallows. They're entangled with green algae whose surface is studded with tiny bubbles of oxygen. As I watch, a bubble shimmers toward the surface and breaks.

The traditional knowledge of the Zuni people tells that the world began as clouds and water until the marriage of earth and sun bought forth green algae. And from the algae there arose all the forms of life. Scientific knowledge tells us that, before the world was green, the only life was in the water. In shallow bays, waves broke on an empty shore. The sunbaked continent was without a single tree to make a pool of shade. The early atmosphere had no ozone, and the sun's full intensity beat down on the land, a deadly rain of ultraviolet radiation, damaging the DNA of any living thing that ventured up on the shore.

But, in the sea and inland ponds where water screened out the UV rays, algae were busily changing the course of evolutionary history, as the Zuni story explains. Oxygen bubbled from the algal strands, the exhaust fumes of photosynthesis accumulating molecule by molecule in the atmosphere. Oxygen, this new presence, reacted with strong sunlight in the stratosphere to produce the ozone layer that one day would shelter all terrestrial life under its umbrella. Only then did the surface of the land become safe for the emergence of life.

Freshwater ponds provided easy living for green algae. Supported by the water itself and constantly bathed in nutrients, the algae had no need for complex structure, no roots, no leaves, no flowers, simply a tangle of filaments to catch the sun. Sex in this warm bath was easy and uncomplicated. Eggs released from slippery strands floated aimlessly about, and sperm were released freely into the water. New algae would grow from that chance fusion of egg and sperm without need of a protective womb, the water providing everything.

Who knows how it happened, the migration from the easy life in the water to the rigors of the land? Maybe the pools dried up, leaving algae stranded on the bottom like fish out of water. Maybe algae colonized the shady crevices of the rocky shore. Fossils record successful outcomes and rarely preserve the process. But we do know that during the Devonian era, 350 million years ago, the most primitive land plants ever seen emerged from the water to try and make a living on the land. These pioneers were the mosses.

To leave this easy aquatic life behind and venture out onto the land posed formidable challenges, chief among them the matter of sex. The algal ancestors handed down the legacy of floating eggs and swimming sperm, which was fine in the water, but a liability on dry land. A drying pond would be the end of peeper eggs. The drying air would doom an alga egg, as well. The life cycle of mosses evolved to meet these challenges.

Once my basket is full of greens, I take out an old canning jar and scoop it full of pond water and peeper eggs. I'll bring them back for my girls to watch as the eggs change into tadpoles. This fascinated me as

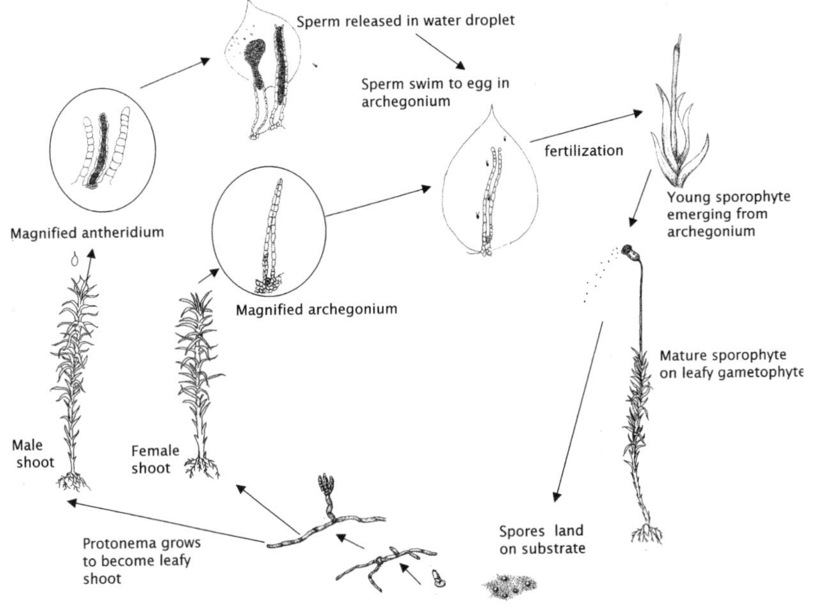

Life cycle of a typical moss

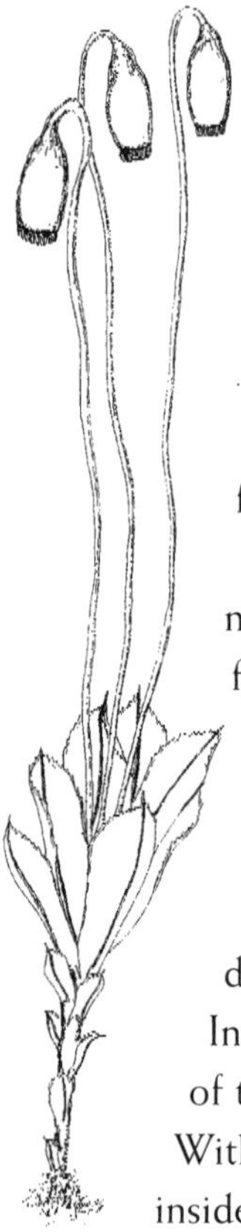

a kid, seeing the black speck in the middle of the egg sprout legs and a tail. The fat, rounded eggs remind me of being pregnant, the sensation of carrying around my own wiggling tadpole in the warm pond within. Each in our own way, we all go back to the pond to reproduce, connecting to our watery origins. The pond bank is tufted with mosses, so I take a clump of them, too. I can stick the moss under the microscope for the kids to look at.

In order to survive on land, mosses evolved a whole new architecture that surpassed the simpler algae. The floating ribbons of algae were replaced by stems that can hold themselves upright. Under the microscope you can see the whorls of perfect little leaves and tiny root-like rhizoids, a tuft of brown fuzz that anchors them to the soil. At the tip of the shoot, the leaves look different and are all clustered together in a tight circle. Invisible, concealed among bunched-up leaves at the tip of the moss, lies the female structure, the archegonium. With gentle probing I can part the leaves to see what is inside. Nestled among them are three or four structures, chestnut brown and shaped like long-necked wine bottles. On another stem, in the axil of a leaf, is another tuft of hair-like leaves. Pushing them aside, I find a cluster of sausage-shaped sacs, each one green and bulging. These are the antheridia, the male structures turgid with sperm ready to be released.

Mosses made a huge innovation to cope with the difficulties of reproduction on dry land. The egg is protected inside the female, rather than cast out on the water. All contemporary plants, from ferns to fir trees, use this same strategy that first arose in the mosses. Like a protective womb, the swollen base of the archegonium holds the egg. The clustered leaves trap water, keeping the egg from drying out and creating a pool through which the sperm can swim. The unfertilized egg sits safely in the archegonium and waits.

But getting sperm to egg is an exceedingly difficult task. The first obstacle is simply the necessity for water, which is unreliable in the terrestrial world. In order to get to the egg, swimming sperm must have a continuous film of water. Rain and dew are captured between the densely packed leaves. The capillary spaces between leaves channel the water among plants, forming a transparent aqueduct between male and female. But any break in the water film poses an uncrossable barrier that prevents the sperm from reaching the egg. It's a race between the sperm and evaporation, which will steal away the temporary bridges of water. Unless the moss is soaked with rain, dew, or the spray of a waterfall to carry the sperm, the eggs will remain unfertilized. In a dry year, reproduction will likely fail.

Moss sperm are produced in great numbers, but each tiny cell has a vanishingly small chance of ever finding an egg. Unlike the peepers who call so strenuously to their mates, moss sperm have no signal to guide them to their destination and so swim at random in the water film. Most are simply lost in the labyrinth of leaves. The small sperm are weak swimmers and carry limited energy for their travels. Once released from the antheridium, the clock begins to count down their survival probability. Within an hour, all will be dead, having exhausted their resources. The eggs continue to wait.

The third challenge lies in the nature of water itself. From the vantage point of a human, water seems so fluid as we dive easily into its depths. But, at the scale of microscopic moss sperm, getting through water is like a person trying to swim through a pool filled with Jell-O. The surface tension of a water droplet presents an elastic barrier for moss sperm; despite wiggling and pressing against it, they can't break through. However, they have devised a number of ingenious means of escaping the grip of water. When the sperm are ready to be released, the antheridium absorbs excess water, swelling until it bursts. The sperm are pushed out under hydraulic pressure, given a head start on their journey.

Another way the moss overcomes the surface tension of water is by packaging a surfactant along with the sperm. When the antheridium ruptures, the chemical surfactant acts like soap, making water less viscous. As soon as the surfactant meets a taut droplet of water, the surface tension breaks and the dome of the drop suddenly flattens to a moving sheet of water, carrying the sperm along like surfers on a wave.

The sperm need all the help they can get in moving toward the egg, seldom travelling more than four inches from the antheridium that produced them. Some species have devised other means to increase the distance, harnessing the power of a splash to spread the sperm. In species like *Polytrichum*, the antheridia are surrounded by a flat disk made of leaves, radiating around them like the petals of a sunflower. A raindrop plummeting onto this disk can splash the sperm as far as ten inches, more than doubling the distance they can travel.

If all the conditions are right, the sperm will be able to swim to the female and down the long neck of the archegonium to the waiting egg. Fertilization yields the first cell of the next generation, the sporophyte. In the life of spring peepers, the fertilized eggs are at the mercy of the environment, floating in the pond protected only by a gelatinous coating. But the moss mothers don't abandon their young. They nurture the next generation right inside the archegonium. Special transfer cells, akin to those in a placenta, allow nutrients to flow from the parent moss to the developing offspring. How amazing to have such kinship with these plants, who nourish their children with cells so like the ones that helped bring my daughters into the world.

The fertilized peeper eggs transform first to tadpoles and then to copies of their parents. Young mosses don't grow directly into adults, leafy copies of their parents. Instead, the fertilized egg develops into an intervening generation, the sporophyte. Still attached to the parent, and nourished by it, the sporophyte will create and disperse the next generation.

Back at the pond, the summer has warmed the water and my daughters and I are tempted by a swim. But the water is murky with algae and none too inviting, even on this hot day. So we stretch out on our bellies on the bank and soak up the sun instead, our books open on the ground. I like having the ground at eye level. I idly run my fingertip over the sporophytes that have formed on the mosses along the shore. Resilient, they spring back from my touch with a little puff of spores released on the breeze. Each one arises from the stem tip, where the archegonium had sheltered the egg in the spring. The sporophyte is now a plump, barrel-shaped capsule at the end of an inch-long stalk, the seta. Inside are masses of powdery spores to be sent off to seek their fortune wherever the wind may carry them.

Finding a home is an improbable business; most spores will haphazardly land in unsuitable places. But should a spore drift to the ground at the moist edge of another pond or in another suitable, moist place, another transformation will take place. The round amber spore will swell with moisture and send out a green thread, the protonema. The threads will branch and spread over the moist ground, establishing a web of green. At this point in its life, the moss most resembles its distant relative: it is nearly indistinguishable from filamentous green algae. Like a newborn bearing the face of great-grandma, the protonema has all the attributes of its algal ancestor, an evolutionary echo contained in its genes. But the resemblance soon passes and leafy shoots rise up from buds along the protonema and form a new thick moss turf.

Most moss stories don't have such a happy ending. Mosses are amateurs in the business of reproduction on land, and it shows. Their adaptations allow them to reproduce, but with very low efficiency. Few sperm ever make it to the vicinity of the archegonium, and many eggs are left waiting like disappointed brides at the altar, an enormous waste of energy. So many factors conspire against successful sexual reproduction, is it any wonder that many species of mosses have given up on sex altogether? Production of sporophytes is rare for many species and totally unknown for others.

Without sexual reproduction, there would be no more peepers, no ringing chorus in the spring. But unlike peepers, mosses can still spread and multiply, even when sperm never meets egg. Sex is not the only opportunity for propagating themselves. Long before the advent of biotechnology, mosses have been making clones, saturating the environment with genetically identical copies of themselves. In fact, most species of moss can regenerate themselves entirely from just a small fragment. A single leaf, broken off by accident and lying on moist soil, can produce an entire new plant. Asexual propagation can also be an alternative reproductive plan. Gemmae, bulbils, brood bodies, branchlets—mosses produce a whole menu of specialized asexual propagules, which are simply detachable appendages on various parts of the moss plant. They break off and disperse to new habitats, where they can form new colonies without the costs and inefficiency of sex. Cloning eliminates the need for getting egg and sperm together, and spending time and energy to produce a sporophyte. All these ways of going on in the world, sexual and asexual, are a complex dance of genes and environment, evolutionary variations on the theme of continuity, of perpetuation.

Every spring, my daughters and I say that the peepers are "calling up the daffodils" with their song. The green shoots push up after the first peepers are heard and come to full flowering before they are finished. My Potawatomi ancestors had a word for this mystery: *puhpowee*, the power that causes a mushroom to rise up from the earth overnight. I think it is this that draws me to the pond on a night in April, bearing witness to *puhpowee*. Tadpoles and spores, egg and sperm, mine and yours, mosses and peepers—we are all connected by our common understanding of the calls filling the night at the start of spring. It is the wordless voice of longing that resonates within us, the longing to continue, to participate in the sacred life of the world.

Sexual Asymmetry and the Satellite Sisters

Our local NPR station has a show in the Saturday morning lineup that usually accompanies my Saturday errands or a drive to the mountains. Sandwiched in between "Car Talk" and "What Do You Know?" is "The Satellite Sisters." "We're five sisters on two continents, with the same parents and living very different lives. Let's talk." The sisters check in by telephone from all over the world, but the show has the feel of a kitchen table, half-filled coffee cups, and a plate of sticky buns. The talk wanders from career strategies to kids, women as environmental activists, and the ethical dilemma posed by sampling grapes in the grocery store. And of course, relationships.

My husband is home puttering in the barn, my daughters are at a birthday party, and I'm feeling as contented and lazy as the Sisters' conversation this morning. Too wet to walk, too muddy to garden, the morning is mine, all mine, and I've been wanting to take a look at all these unidentified *Dicranums*. What luxury, to come to work in order to play. The rain streams down the windows of the lab and the voices of the Sisters are my only company. I can laugh out loud with them, and who will notice? There are no students, no phone calls, just some handfuls of moss and a few hours stolen from the normal hubbub of a weekend.

Dicranum is a genus of moss with many species, sisters in the same family. I think of them as only females, since the menfolk have met an unusual, and perhaps fitting, fate which strong women will understand immediately. We'll get to that later. While the Sisters trade stories of the vulnerability that comes with a new hairstyle, the exposure of a tentative self, I'm laughing at how I'd never noticed before that the *Dicranums*, more than any other moss, look like hair, combed hair, neatly parted and swept to one side. Other mosses bring to mind carpets or miniature forests, but *Dicranum* evokes hairstyles: ducktails, waves, corkscrew curls, and buzzcuts. If you lined them up for a family photograph, from

Dicranum montanum

the smallest, *D. montanum*, to the largest, *D. undulatum*, you'd definitely see the family resemblance. They all have the same hair-like leaves, long and fine with a curl at the end, brushed in one direction for that windswept look.

Like the Satellite Sisters phoning in from Thailand and Portland, Oregon, the *Dicranums* are widely distributed in forests all over the world. *Dicranum fuscesens* lives in the far north, while *D. albidum* goes all the way to the tropics. Perhaps the distance between them helps for peaceful coexistence between siblings. The genus *Dicranum* has undergone considerable adaptive radiation, that is, the evolution of many new species from a common ancestor. Adaptive radiation, whether in Darwin's finches or in *Dicranum*, creates new species that are well adapted for specific ecological niches. Darwin's finches evolved from a single ancestral species lost at sea and swept out to the barren Galapagos Islands, where the birds evolved into new species. Each island in the archipelago supports a unique species, with a unique diet. Likewise, the original *Dicranum* diverged into many different species, each with a distinctive appearance and habitat, lifestyle variations on the ancestral theme.

The force behind this divergence into new species is related to the inevitable competition between siblings. Remember wanting what your brother had, just because he had it? At the family dinner table, if everyone wants a drumstick from the Sunday chicken someone will be disappointed. When two closely related species put the same demands on their environment, with not quite enough to go around, both will end up with less than they need to survive. So, in families, siblings can coexist by developing their own preferences, and if you specialize in white meat or the mashed potatoes, you can avoid competition for the drumsticks. The same specialization has taken place in *Dicranum*. By sidestepping competition, numerous species can coexist, each in a habitat that they don't have to share with a sibling species, the mosses' equivalent of "A Room of One's Own."

In the *Dicranum* clan, there are family roles that could easily apply to sisters in any big family. You'll recognize them right away. *D. montanum* is

Dicranum scoparium

the unassuming one; you know the type—nondescript, overlooked, with her short curls always in disarray. She's the one who gets the leftover habitats, the chicken wings of the Sunday feast, the occasional exposed root of a tree or bare rock. Moist shady rocks are also the habitat of the glamorous *D. scoparium*, the one who draws the looks with long, shiny leaves, tossed to one side. This is the plush *Dicranum*, the one that makes you want to run your hand over its silkiness and pillow your head in its deep cushions. When these sister species live together on a boulder, the showy *D. scoparium* takes all the best places, the moist sunlit tops, the fertile soil, while *D. montanum* fills in the gaps. No one is surprised when *D. scoparium* crowds out the little sister, invading her space and driving her to the edges.

The other *Dicranums* tend to avoid the conflicts that arise from sharing the same space, where strong identities can clash. *D. flagellare*, with leaves trim and straight, like a military buzzcut, remains aloof from the others, choosing to live only on logs in an advanced state of decay. She's the conservative one, celibate for the most part, foregoing family in favor of her own personal advancement by cloning. Solitary and intensely green, *D. viride* has a hidden fragile side, with leaf tips always broken off like bitten fingernails. *Dicranum polysetum*, on the other hand, is the most prolific mother of the family, an inevitable outcome of her multiple sporophytes. Then there's the long, wavy-leafed *D. undulatum*, capping the tops of boggy hummocks, and *D. fulvum*, the black sheep of the family; more than a dozen species of powerful females.

I'm filling a second cup of coffee and patiently cataloging the moss samples, when the Saturday conversation of the Satellite Sisters wanders to men. Some of the sisters are happily married and others are sharing last weekend's episode of looking for Mr. Right, pondering commitment and probable fatherhood personalities. Finding the right mate, a universal female concern, is also an issue for *Dicranum*. Sexual reproduction in mosses is an iffy business, as we have seen, given the limited abilities of the weak, short-lived males. Thwarted by a lack of swimmable water between them and the egg, their success depends on well-timed rainfalls. The sperm must swim to the egg, facing barriers that isolate them even though they are only a few inches apart. So near and yet so far, most eggs sit and wait in the archegonium for a sperm that never comes.

Some species have evolved a means to increase their chances of finding a mate. They become bisexual. After all, fertilization is virtually guaranteed when egg and sperm are produced by the same plant. The good news is that there will be offspring; the bad news, they are all inbred. None of the *Dicranum* species have evolved the bisexual lifestyle, keeping distinctions between the genders very clear indeed.

Dicranum fulvum

Given the difficulties in getting males and females together, it's surprising how common it is to see a colony of *Dicranum* bristling with sporophytes, the outcome of numerous sexual encounters. I have a clump of *D. scoparium* here, which must have fifty sporophytes on it, representing potentially fifty million spores. How do they do it? You might guess that the key to their reproductive success was a very favorable sex ratio, with numerous males hovering around every female. Some mosses have adopted this strategy, but not *Dicranum*.

While the radio Sisters compare their rules for first dates, I take this clump of *Dicranum* apart, looking for the macho males who are responsible for all these babies. The first shoot I pull out is a female. So is the second. And the third. Every single shoot in the colony is a female, and yet every single shoot has been fertilized. Pregnant females without a male in sight? Immaculate conception has not yet been documented in mosses, but it makes you wonder.

I slide one of the female shoots under my microscope for a closer look, and see just what I'd expect: the female anatomy, fertilized eggs swelling with the next generation. The stem is covered by a sheaf of long leaves, swept gracefully to the side with that unmistakable *Dicranum* flair. I follow one of the curving leaves along its arc, its smooth cells and shining midrib. And then I notice a whiskery little outgrowth, something I've seen only once before. Dialing up the magnification of my microscope, I can see that it's a tiny little cluster of hair-like leaves, a miniature plant growing from the massive *Dicranum* leaf like a clump of ferns growing on a tree branch. At even higher power, sausage-shaped sacs come into focus, unmistakably antheridia, swollen with sperm. Here are the missing fathers: microscopic males reduced to hiding out among the leaves of their would-be mates. They have entered female territory with a single purpose, a kind of stealthy intimacy, putting themselves so close to the females that the impotent sperm easily swim the distance to the egg.

Dicranum polysetum

Females dominate all aspects of *Dicranum* life, in numbers, in size, in energy. Whether or not males even exist lies in the power of the females. When a fertilized female produces spores, those spores are without gender. Each one is capable of becoming a male or a female, depending on where it lands. If a spore drifts to a new rock or log that is unoccupied, it will germinate and grow up to be a new full-size female. But should that spore fall onto a patch of *Dicranum* of the same species, it will sift down among the leaves of the existing females and become trapped there, where the female will control its fate. The female emits a flow of hormones which cause that undecided spore to develop into a dwarf male, a captive mate that will become the father of the next generation in the matriarchy.

The Sisters are interviewing someone on the effects of the two-career family. I want to call in to the show to see what they'd say about *Dicranum's* domestic arrangement. Five sisters, five perspectives on dwarf males: a clear case of female tyranny, surrender of masculinity to strong women, turnabout is fair play ... hey, give them the benefit of the doubt, it's possible that they are sensitive '90s kinds of guys, giving the females their space. Will they still think that size matters?

In this time and place, men and women have the luxury of creating our relationships quite independently of their survival value to our species. Heaven knows there are plenty of us already. The ways we negotiate the balance of power and domestic harmony are unlikely to change the trajectory of our population.

But from the evolutionary perspective of *Dicranum*, the asymmetry of the sexual relationship matters a great deal. Dwarf males are an efficient solution to the problem of getting fertilized. The entire species, both sexes, benefits from this arrangement. A full-size male actually stands in the way of his own genetic success, his leaves and branches increasing the distance between sperm and egg. A dwarf male will produce many more offspring than will a full-size male. He can best contribute to the next generation by delivering sperm and then getting out of the way.

The very same impulse that propels sister species to diverge from one another creates the sharp difference between *Dicranum* males and females. Competition in a family decreases everyone's potential success. So evolution favors specialization, avoiding competition, and thus increasing the survival of the species. A large female and a dwarf male cannot compete with each other. The male is small, to better deliver sperm. The female is large to nurture the resulting sporophyte, their future, their offspring. Without competition from their mates, females get all of the good habitat, the light and water and space and nutrients, all for the benefit of the offspring.

The hour with the Satellite Sisters is winding down, with a recipe for lemon mousse. It sounds great. The rain has stopped and my mosses are done, so I turn off the radio with a smile. It's time to go home for lunch, lovingly prepared by my full-size male.

An Affinity for Water

O n the hilltops of my home in upstate New York, the bare gray branches of the maples seem to be traced with a newly sharpened pencil against the winter sky. But in the Willamette Valley the Oregon oaks are drawn in thick green crayon. The steady winter rains keep the tree trunks lush and green with moss, while their leaves lie dormant. This mossy sponge drips a constant flow of water to the tree roots, saturating the ground below and filling the soil reservoir for the summer ahead.

By August, the winter rains have long been consumed, and the land is thirsty again. The oak leaves hang in the hot air and the buzzing cicadas broadcast the weather forecast: the 65th day without rain. The wildflowers have retreated underground to avoid the drought, leaving a landscape of parched brown grasses. The moss carpets now lie desiccated on the bark of the summer oaks, their shriveled, wiry skeletons barely recognizable. In the summer drought, the oak grove is hushed and waiting. All growth and activity is suspended in drought-sleep.

Linden's plane is late, so I amble down to the AeroJava stand, joining the line, killing time. There is a jar on the counter, half full of dimes and pennies, with a hand-lettered sign: "If you fear change, leave it here." Unaccountably, I find my eyes filling for a moment, wishing I could empty my pockets, casting off my load of change, bringing my daughter back, the little one, standing on a chair with my apron tied three times around her, cutting out Valentine cookies and splattering the kitchen with pink frosting.

The mosses begin their time of waiting. It may be only a matter of days before the dew returns, or it may be months of patient desiccation. Acceptance is their way of being. They earn their freedom from the pain of change by total surrender to the ways of rain.

I've lost a lot of days to waiting, holding my breath until circumstances have changed, straining toward the scent of rain. I remember waiting for what seemed like forever to be big enough to ride the school bus, which gave way to waiting for that same bus, stomping my feet against the biting cold. Waiting nine sweetly rounded months for the arrival of my babies was followed too soon by waiting for them outside the high school basketball game, fingers tapping impatiently on the steering wheel. And now I wait for the touchdown of Linden's plane, bringing her back from college, waiting to slip my arm through hers, while we wait together at my grandfather's bedside.

What art of waiting is practiced by the mosses, crisped and baking on the summer oak? They curl inward upon themselves, as if suspended in daydreams. And if mosses dream, I suspect they dream of rain.

Mosses must be awash in moisture in order for the alchemy of photosynthesis to occur. A thin film of water over the moss leaf is the gateway for carbon dioxide to dissolve and enter the leaf, beginning the transformation of light and air into sugar. Without water a dry moss is incapable of growth. Lacking roots, mosses can't replenish their supply of water from the soil, and survive only at the mercy of rainfall. Mosses are therefore most abundant in consistently moist places, such as the spray zone of waterfalls and cliffs seeping with spring water.

But mosses also inhabit places that dry out, such as rocks exposed to the noonday sun, xeric sand dunes, and even deserts. The branches of a tree can be a desert in the summer and a river in the spring. Only plants that can tolerate this polarity can survive here. The bark of these Oregon oaks is shaggy with *Dendroalsia abietinum* all year round. The name *Dendroalsia* translates from scientific Latin to something like "Companion of Trees." Like others of its kind, beautiful *Dendroalsia* tolerates these wide swings in moisture, with a suite of evolved adaptations known as *poikilohydry*. Its life is tied to the comings and goings of water. Poikilohydric plants are remarkable in that the water content of the plant changes with the water content of the environment. When moisture is plentiful, the moss soaks up the water and grows prolifically. But when the air dries, the moss dries with it, eventually becoming completely desiccated.

Dry curled shoot of Dendroalsia

Such dramatic drying would be fatal to higher plants, which must maintain a fairly constant water content. Their roots, vascular systems, and sophisticated water-conservation mechanisms allow them to resist drying and stay active. Higher plants devote much of their effort to resisting water loss. But when water depletion becomes severe, even these mechanisms are overcome, and the plants wilt and die, like the herbs on my windowsill when I left for vacation. But most mosses are immune to death by drying. For them, desiccation is simply a temporary interruption in life. Mosses may lose up to 98 percent of their moisture, and still survive to restore themselves when water is replenished. Even after forty years of dehydration in a musty specimen cabinet, mosses have been fully revived after a dunk in a Petri dish. Mosses have a covenant with change; their destiny is linked to the vagaries of rain. They shrink and shrivel while carefully laying the groundwork of their own renewal. They give me faith.

Linden steps off the plane so glad to be home, beaming a girl's smile, but her woman's eyes scan my face for signs of concern. I grin back reassuringly and hug her tight. Walking along beside her I see right away that she has not been wasting her days in waiting; she has been becoming. I know now there's nothing in the world that would have me trade this lovely young woman, radiant, with her arm linked through mine, for the toddler who slept in my arms.

Moistened shoot of Dendroalsia with sporophytes

Poikilohydry enables mosses to persist in water-stressed habitats which more advanced plants cannot endure. But this tolerance comes at a serious cost. Whenever the moss is dry, it cannot photosynthesize, so growth is limited to brief windows of opportunity when the moss is both wet and illuminated. Evolution has favored those mosses that can prolong this window of opportunity. They have elegantly simple means of holding on to precious moisture. And yet, when the inevitable drought arrives, their acceptance is total and they are beautifully equipped for endurance, waiting until the rains return.

The atmosphere is possessive of its water. While the clouds are generous with their rain, the sky always calls it back again with the inexorable pull of evaporation. The moss isn't helpless; it exerts its own pull to counter the powerful draw of the sun. Like a jealous lover, the moss has ways to heighten the attachments of water to itself and invites it to linger, just a little longer. Every element of a moss is designed for its affinity for water. From the shape of the moss clump to the spacing of leaves along a branch, down to the microscopic surface of the smallest leaf; all have been shaped by the evolutionary imperative to hold water. Moss plants almost never occur singly, but in colonies packed as dense as an August cornfield. The nearness of others with shoots and leaves intertwined creates a porous network of leaf and space which holds water like a sponge. The more tightly packed the shoots, the greater the water-holding capacity. A dense turf of a drought-tolerant moss may exceed three hundred stems per square inch. Separated from the rest of a clump, an individual moss shoot dries immediately.

I feel myself expand in her presence. Her stories make me laugh and waken my own stories to intertwine with hers. With her next to me in the car, fiddling with the radio to find her favorite station, I somehow know myself again, and know that the ache her absence brings is not only about losing her, but about losing it all, my grandfather, my parents, myself. How fearfully we fight the losses that Dendroalsia *so gracefully embraces. Straining against the inevitable, we spend ourselves on futile resistance, as if we could somehow outrun the drying of a dewy cheek.*

Water has a strong attraction for the small spaces in a clump of moss. The water molecules attach themselves readily to leaf surfaces, due to the adhesive properties of water. One end of the water molecule is positively charged, the other is negative. This allows water to adhere to any charged surface, positive or negative, and the moss cell wall has both. The bipolar nature of water also makes it cohesive, sticking to itself, with the positive end of one molecule linking up head to tail with the negative end of another. As a result of this strong cohesion and adhesion, water can form a transparent bridge between two plant surfaces. The tensile strength of this bridge is sufficient to span small spaces, but collapses if the gulf is too broad. The delicate leaves and small stature of mosses provide spaces of just the right size for these bridges to form by the capillary forces of water. Moss shoots, branches, and leaves are arranged in such a way as to prolong the residence time of water and to counter the forces of evaporation, with the pull of capillarity. Mosses without such favorable design dried out too quickly and were eliminated by natural selection.

Watch a drop of rainwater fall on a broad, flat oak leaf. It beads up for a minute, reflecting the sky like a crystal ball, and then slides off to the ground. Most tree leaves are designed to shed water, leaving the task of water absorption to the roots. Tree leaves are covered with a thin layer of wax, a barrier to water entering by absorption or leaving via evaporation. But moss leaves have no barrier at all, and are only one cell thick. Every cell of every leaf is in intimate contact with the atmosphere, so that a raindrop soaks immediately into the cell.

On the way to the hospital, we talk and talk, sometimes about her great-grandpa, but mostly about an amazing slice of time, her freshman year in progress. She tells me about her classes, people I've never met, a backpacking trip—I hear passions she never guessed at, courageous ventures into unknown territory. Listening, I realize I'm a little envious of her openness to the world, where change is only the lure of imagined possibilities, not the agent of impending losses. But I know there is no barrier I can construct which will hold those losses back, that won't also shut me out, alone, and disconnected from the world.

The leaves of trees are uniformly flat, to intercept as much light as possible, and spaced far from one another to prevent shading. But light is of less concern to mosses than is water. Therefore, the nature of moss leaves is entirely different from trees'. Each leaf is shaped to make a home for water. Lacking roots or an internal transport system of any kind, mosses rely entirely on the shape of their outer surfaces to move water. In some species, the flow of water is accelerated by the wicking action of minute threads, or *paraphyllia*, that densely cover the moss stem, like a blanket of coarse wool. The shape and arrangement of some moss leaves collect and retain water, a concave leaf holding a single raindrop in its upside-down bowl. Others have long leaf tips, rolled into tiny tubes that fill with water and channel the droplets to the leaf surface. Leaf overlaps leaf, closely spaced, creating tiny concave pockets, a continuous conduit for water moving among them.

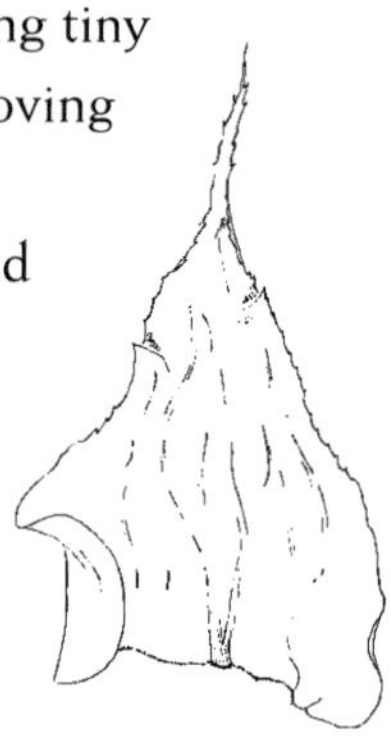

Even the microscopic surface of the leaf is sculpted to attract and hold a thin film of water. The leaves may be pleated into minute accordion folds that trap water in the crevices, undulations in the leaf creating a microtopography of rolling hills and water-filled valleys. Species of arid habitats often have leaf cells that are studded with tiny bumps called *papillae*, forming a roughened surface you can feel if you rub the leaf gently between your fingertips. A film of water stretches between the papillae, which rise like little hills above a lake, allowing the leaf to hold water and photosynthesize just a little longer, even while the sun beats down.

The top shelves of my office are stacked with cartons of dry mosses, packed away as references to one research project or another. Each time I take out a specimen it must be wetted so that I can see its fine-scale features on which identification depends. I suppose I could just soak it in a Petri dish for a few minutes. But even after all these years, I still delight in the ritual of adding the water, drop by drop, and watching with the microscope as the shoots revive. I think of it as a small act of homage to a remarkable marriage between moss and water. The moss and the water seem to have a magnetic attraction for each other. I add a drop of water to the tip of the dry shoot, and it rushes among the moss leaves like a flash flood down a narrow canyon. Dry contorted leaves unfurl, all is light and movement as the drops follow every passageway and penetrate every little space, swelling beneath the convex leaves and bowing them outward.

Just where the leaf attaches to the stem are specialized groups of alar cells. To the naked eye, they appear as shiny crescents at the corners of the leaves. Under the microscope, alar cells are much larger than a typical leaf cell and often are thin walled. The large empty space of the alar cell absorbs water rapidly and can inflate like a transparent water balloon. This swelling causes the leaf to bend out and away from the stem, to a more favorable position to capture light. Without nerve or muscle, the moss can sense the water that makes growth possible, and adjust the leaf angle to the optimal plane for photosynthesis. Leaf bases fill and overflow, with the excess drawn to the leaf below, creating an interconnected string of pools beneath each overlapping leaf. Within minutes the shoot is saturated and the water comes to rest, leaving the shoot plump and shining. And then it's over. The shape of the water is changed by the moss and the moss is shaped by the water.

The mutuality of moss and water. Isn't this the way we love, the way love propels our own unfolding? We are shaped by our affinity for love, expanded by its presence and shrunken by its lack.

Plants and animals of all kinds have sophisticated means of maintaining water balance, using pumps and vessels, sweat glands and kidneys. Considerable energy is devoted to water regulation in these organisms. But mosses engineer the movement of water simply by harnessing the attraction of water for surfaces. Their forms take advantage of the adhesive and cohesive forces of water, to move the water at will over their surfaces, without expending any energy of their own. This elegant design is a paragon of minimalism, enlisting the fundamental forces of nature, rather than trying to overcome them.

My grandfather would have appreciated the elegant design of a moss, if he'd ever had the chance to see it. He was a carpenter. His shop was a warren of tools; precision lathes and hand drills, antique planes and sculpting chisels; each tool to its own purpose. No material was wasted; there were baby-food jars of carefully sorted screws, a walnut board, a salvaged oak newel post waiting to be transformed into a bowl for my grandmother's kitchen. His designs were clean and simple, to match the potential of the wood to the task at hand.

Despite all these remarkable tactics for water retention, they are only a temporary respite from evaporation. The sun always wins the battle and the moss begins to dry. Profound changes in the shape of the mosses occur as water is pulled back to the atmosphere. Some mosses begin to fold their leaves, or roll them inward. This reduces the exposed surface area of the leaf and helps the plant cling to the last bits of surface water. Nearly all mosses change their shape and color when they dry out, making identification of species doubly difficult. Some leaves wrinkle, some spiral and twist their leaves around the stem, a sheltering cloak against the dry wind. The plumes of *Dendroalsia* darken and coil inward looking like the black tail of a mummified monkey. Crisp, dry, and contorted, the mosses are transformed from soft fronds to brittle, blackish tufts.

My grandfather looks too tall for his hospital bed, surrounded by the thicket of equipment that is keeping him alive. His softness seems like an alien presence in this realm of hard surfaces, sharp angles, and the confident hum of electronics. An IV tube runs into his arm, battling dehydration. It's calibrated to maintain the 87% of his body that is water, while the other 13% begins the march of surrender.

At the same time that drought begins to shrink the moist leaves, preparations for drying are also going on in the biochemistry of the moss cell. Like a ship being readied for dry dock, the essential functions are carefully shut down and packed away. The cell membrane undergoes a change that allows it to shrink and collapse without sustaining irreparable damage. Most importantly, the enzymes of cell repair are synthesized and stored for future access. Held in the shrunken membrane, these lifeboat enzymes can restore the cell to full vigor when the rain returns. The internal machinery of the cell can turn on and quickly repair the desiccation damage. Only twenty minutes after wetting, the moss can go from dehydration to full vigor.

We stand together in the cemetery with all the paraphernalia of resistance now set aside. With her hand in mine, my grandmother's face is brittle and ready to break. My mother's gaze moves among us, gathering each of us in. My pink-cheeked child shifts from foot to foot, not knowing where to stand. She stands in a circle of daughters, with hands joined, where one day she will be the one to let go. When the roses slip from her hands, we hold each other's tighter.

Holding water against the pull of the sun, and welcoming it back again is a communal activity. No moss can do it alone. It requires the interweaving of shoots and branches, standing together to create a place for water.

The soft fall clouds finally darken the hard summer sky and a wet wind stirs the dry oak leaves scattered on the ground. The air is charged with energy, as if the mosses are poised and alert, tasting the wind for the scent of rain. Like captives of the drought, their senses are tuned to the approach of their rescuers.

When the first drops begin to fall, and shower turns to downpour, it is an exuberant reunion. The water courses through the old paths constructed especially to welcome its arrival. Flooding down the canals of tiny leaves, the water finds its way through the capillary spaces and soaks deeply into every cell. Within seconds the eager cells grow turgid and contorted stems spread toward the sky, leaves outstretched to meet the rain. I run out to the grove when the rains arrive, I want to be there when the unfolding begins. I, too, can have a covenant with change, a pledge to let go, laying aside resistance for the promise of becoming.

Animated, released from stillness by the rain, *Dendroalsia* begins to move, branch by delicate branch unfolding to recreate the symmetry of overlapping fronds. As each stem uncurls, its tender center is exposed and all along the midline are tiny capsules, bursting with spores. Ready for rain, they release their daughters upon the updrafts of rising mist. The oaks once more are lush and green and the air smells rich with the breath of mosses.

Binding up the Wounds: Mosses in Ecological Succession

In the after-lunch indolence that follows a good climb, I'm watching an ant haul a sesame seed from my sandwich crumbs across the bare rock. She carries it to a crevice in the rock which is filled with *Polytrichum*, a bristly moss that has taken hold in the tiny pocket of soil collected there. I doubt that next summer's hikers will find a sesame sprout, but the crack already holds a tiny spruce, started from a seed among the mosses. The ant, the seed, the moss, all bent on their own paths, are unwittingly working together, covering open ground, sowing a forest on this bare rock. The process of ecological succession is like a positive feedback loop, a magnet of life attracting more life.

From the dome of Cat Mountain, the Five Ponds Wilderness stretches out at my feet, the largest wilderness area east of the Mississippi, rolling green hills stretching to the horizon. This sun-warmed granite is some of the oldest rock on Earth, and yet the forest below is relatively new. Only a century ago the redtails would have ridden the thermals over charred ridgetops, cutover valleys, and isolated pockets of old-growth forest. The Adirondacks have been called "The Second Chance Wilderness." Today bears and eagles fish along the meandering course of the wild Oswegatchie River. Its logging scars healed by succession, it is an unbroken expanse of second-growth forest. Unbroken, save for one open wound. To the north, the green is interrupted by a gash, a treeless barren visible from ten miles away.

The rock hereabouts is rich in iron. There are places where your spinning compass makes you think you've wandered into the twilight zone. You can pick up the beach sand with a magnet. Iron mining came early to the Adirondacks and at Benson Mines they tore down the mountain and ground it up. The ore went all over the world, and the mountain came out of a pipe as a slurry of tailings, mine waste laid thirty feet deep. Then the bottom dropped out of the market, jobs left, the mine closed and left behind hundreds of acres of sandy waste, a Sahara in the midst of the wet green Adirondacks.

Current law demands restoration of mined lands, but Benson Mines fell between the legal cracks and has not been reclaimed. There were some half-hearted attempts at revegetation, but they all failed. Midwestern prairie grasses were planted in some places, but they couldn't survive long without fertilizer and irrigation, which ran out about the same time that the business moved overseas. Someone planted trees; a few of the pines have persisted, yellow and stunted. I don't know whether these were planted as acts of contrition or a façade of responsibility, but there wasn't much sense to it, like painting a mural on a condemned building. It's not enough to put plants here; there has to be something to sustain them and the tailings are a far cry from the humus-rich soils that now lie buried under the sterile sand. Now it's officially classified as an "orphan mine." Rarely is official language so direct and evocative—this piece of land is indeed without anyone to care for it.

Driving the Adirondack roads, past glittering lakes and deep woods, you'll scarcely ever see roadside litter. People love this wild place and the care for its well-being is obvious. But where Route 3 cuts through the mined land, plastic bags are caught in the alders and beer cans float in the ditches full of rusty water. Disregard is also a positive feedback loop; garbage attracts garbage.

I turn in to the cemetery, an anomalous patch of green surrounded by the old mines. The company showed as little regard for the dead as for the living. Past the well-tended gravestones, the pavement ends and the tailings begin. The polished granite monuments give way to an idiosyncratic collection of homemade memorials: a rusty blade from the sawmill half-buried in the ground, rebar welded to form initials, an old-time TV aerial bent to the shape of a cross. There are stories here, buried in the tailings. The path to the mine runs through the cemetery junkpile, past old Christmas wreaths still on their stands, white plastic baskets of pink plastic flowers, the remains of mourning.

I walk up the tailings slope, slipping backwards in the loose sand, like walking on the beach. I don't mind my shoes filling with it; these dunes aren't toxic, just hostile in the way of most deserts. The sand can't hold water so any rains percolate quickly, leaving it dry again. Without vegetation, there's no organic matter to soak up water or build the foundation of a nutrient cycle. Without the shade of trees, the surface temperature can reach extremes—I've measured it at 127 degrees Fahrenheit, more than enough to wither a tender seedling. The slope is littered with spent shotgun cartridges, and cans shot full of holes. And here and there are odd little structures, pieces of fabric stretched between Popsicle sticks, like miniature tents. Pieces of old carpeting lie on the sand, like a strange demonstration from a zealous vacuum cleaner salesman.

Up ahead, I see Aimee kneeling in the tailings, cradling a clipboard, her red curls tucked underneath a broad-brimmed hat. She looks up, apprehensively at first, and then smiles. I know she'll be glad of the help today and relieved to have company. Last week, she found an ugly threat scrawled into the smooth surface of our research plots. Garbage attracts garbage. At least today she'll know that the sound of approaching footsteps is only me.

The role of mosses in ecological succession on the mines is Aimee's thesis project, and she has experiments set up all around. Together, we set off across the tailings to check on some research plots. Where the slope levels out there are tire tracks. Trucks with names like Sippin' Sue and the Honey Wagon painted on the tanks have been making illegal deliveries under the cover of darkness. The stink of septic tank contents lingers in the air. Items people thought they had "disposed of" with a flush now see the light of day again, in a pool of dried-up sewage sludge. The water and the nutrients might actually have done some good if there was soil to hold it. But it all drains away quickly, leaving behind a gray crust studded with cigarette butts and pink tampon applicators. Garbage attracts garbage.

On the other side of the pile, there is a place where the land is healing itself, without benefit of sewage effluent or exotic grasses. Here are clumps of bright hawkweed and clover and scattered evening primroses rooted in the tailings; they would be weeds in other situations, but their presence here is welcome. Especially to the butterflies that crowd around them, as if they were the only flowers around. They are.

Most of this slope is a carpet of *Polytrichum* moss, the same species I saw at the top of Cat Mountain. I admire its tenacity in enduring this place, where others would wither away in the span of a single day. In last year's field season, Aimee found that the wildflowers were almost never rooted in the bare tailings; they almost always occurred in the beds of *Polytrichum*. This summer we're trying to find out how that works. Do the mosses come in under the tiny island of shade produced by the flowers, or do the mosses create a safe place for the weed seeds to get started? How are they interacting to advance succession? She calls me over to a cluster of the little tents that I'd seen on the way up the slope. She erected these canopies to see if moss growth is increased or decreased by shade. Shade might help to explain the association between moss and wildflowers. We kneel and peer under the canopy: the moss beneath it is soft and green, while most of the rest of the slope is black and crisp. Walking over the dry moss sounds like walking over crackers as they break beneath your feet.

I pluck a shoot of *Polytrichum* from under the tent and look at it with my hand lens. The leaves are long and pointed, making the whole plant look like a tiny pine tree. All along the center of each leaf are wavy ridges of bright green cells, the *lamellae*. When the plant is wet, the lamellae are exposed to the sun like solar panels. Like other mosses, it can photosynthesize only when the leaf is both moist and illuminated. Otherwise, which is most of the time, growth is suspended and the moss just waits. It's no wonder that it has taken forty years to carpet this small patch of tailings.

The mossy slope changes color as we work through the day. In the morning light, it's a wash of blue green. The dew of the previous evening was intercepted by the stiff leaf points and funneled down to the leaf base. Moistened, the leaves open and take advantage of the cool morning sun. But when the *Polytrichum* starts to dry, the leaf folds inward to protect the lamellae from desiccation, and growth stops until the next time conditions are right. By lunchtime, the leaves have all flexed inward like a folded umbrella and the green is hidden. Only the dead leaves at the base of the plant are visible, giving the whole slope a black, crusty appearance. With the leaves all folded up, the surface of the tailings is exposed. You have to look closely to see it. When you are down on your knees, the tailings are almost too hot to touch. The surface between the shriveled moss stems is speckled with blackish green. This is a microbial crust, a community even tinier than the mosses, which seem to tower above it. It's made up of intertwined filaments of terrestrial algae, bacteria, and fungal hyphae, taking advantage of the shade provided by the mosses. The algae are nitrogen fixers, incrementally adding nutrients to the tailings.

We try to finish our work by mid-afternoon when its really gets scorching. We can retreat to the shade and get a glass of iced tea at the café in Star Lake, but the *Polytrichum* is stuck out on the tailings in the heat of the day. Its remarkable stress tolerance allows it to persist on this harsh site. It can endure a complete lack of water, while the grasses and wildflowers cannot. *Polytrichum*'s needs for minerals are met by rainwater alone, while the higher plants must extract them from the soil with roots that die in the drought.

The *Polytrichum* carpet is interrupted by small gullies and windswept bare spots. Anyplace the moss is absent, the tailings are subject to erosion. If you pick up a handful of bare tailings, it sifts away through your fingers like water and the wind scatters it as it falls. But under the mosses the tailings are bound firmly together, the sand braided by moss rhizoids. I can stick the blade of my Swiss Army knife into the turf and slice out a neat column of sand, several inches deep, with a cap of moss at the top. The sand below the moss is darkly tinted. This tiny amount of accumulated organic matter may slow the passage of water and subtly increase the pool of soil nutrients. The hair-like rhizoids of *Polytrichum* bind the tailings together and stabilize the surface. We think that this stability might be important as a factor for allowing other plants to get started and Aimee has set up a clever experiment to test that.

It's hard to track the fate of tiny wind-blown seeds that look just like grains of sand. So Aimee went to the bead store and bought vials of plastic beads in the brightest possible colors. Sometimes our kind of science needs more creativity than high-tech equipment. She carefully placed the beads in a grid pattern on different kinds of surfaces at the mine—the bare tailings, under shading vegetation, and on the moss carpet. Every day she would come back and count them. Within two days, every bead on the bare tailings was gone, blown away and buried in the shifting sand. A few more persisted under the wildflowers, but the record setter was *Polytrichum*. The beads became lodged between the shoots, safe from the wind. Mosses might well advance succession simply by providing a safe site for germination. A few days later, a natural experiment confirmed her results. The aspens at the edge of the mine released their cottony cloud of seeds, which blew freely across the bare tailings but were trapped by the moss turf, caught like cat hair on a velvet sofa.

But plastic beads are not seeds, and just because a seed is caught doesn't mean it will germinate and become an established plant. The moss turf is just as likely to hinder the seed as help it, as the two compete for water, space, and scarce nutrients. The moss turf might strand the seed high and dry above the soil and prevent it from ever sprouting or block the passage of tiny roots to the soil. So the next step in our investigation was to sow real seeds. Armed with patience and forceps, Aimee followed the fates of hundreds of seeds, marking each germination and recording its growth over the weeks. In every experiment, with every species, she found that seeds grew and survived best when living in partnership with the moss. The *Polytrichum* seemed to encourage the success of the seedlings. Life attracts life.

Or does it? With appropriate scientific skepticism, we wondered if all the seeds really need is a protective substrate. Maybe it doesn't need to be a living moss at all. The *Polytrichum* may be no more than a physical refuge. How could we see if the seeds were responding just to the protection and not to the moss itself? Can seeds differentiate between moss and a surrogate of the same structure? We puzzled over how to create an experimental substrate which simulated a moss, yet was not alive.

Language provided the key to our experiment. People often refer to mosses as "carpets." The metaphor is really very apt, so we headed for the carpet store. We found ourselves running our hands over Berbers and shags to find the most moss-like textures. Rugs are excellent mimics of the structure of moss colonies, with their closely packed upright shoots. Laughing down the display aisles, we began renaming the designs by their mossy look-alikes: Urban Sophisticate became *Ceratodon*, Country

Tweed was clearly the synthetic kin of *Brachythecium*. We chose a shag carpet that most closely mimicked the *Polytrichum* turf, Deep Elegance. It was wool and so would hold water as well as provide protection. We also bought some remnants of an outdoor carpet, an Astroturf of water-repellent plastic fibers in a lurid shade of grass green. Each piece was subject to abuses never dreamed of in the sales warranty. We soaked it to remove chemicals and punched it full of holes to allow water to percolate through.

We installed the carpet squares out on the tailings with small stakes. Aimee sowed a variety of seeds on each carpet, as well as on the tailings and the living carpet of *Polytrichum*. Faced with a choice between the shag which provided water and shelter, the Astroturf which provided shelter but no water, the real moss, and the bare tailings, what would the seeds do?

Weeks later, the hot summer weather broke with a thunderstorm, echoing off the headwall of the old mine. The desert of the tailings was briefly cooled as the water poured through the sand like a sieve. Unsheltered seeds were washed down the bare gullies. *Polytrichum* unfurled its leaves and began to display resilient green. The Astroturf lay lifeless on the tailings; the shag was sodden and spattered with mud. Out of the carpet of living moss came a crowd of seedlings, the next step in binding up the wounds of the land, life attracting life.

Human communities aren't so different. Like ecological succession, one phase leads to the next. The village at Benson Mines was once a small settlement of loggers in a seemingly boundless forest. Maybe there was just a single house, like a pioneering clump of moss. More families followed, and children, and then a school, a growing population which brought a store and then a railroad and then the mine. It seems that people take as little responsibility for their incrementally evolving future as does a seedling landing on a moss. The corporation left them a legacy, life on the edge of a wasteland, burying their dead in mine tailings.

Aimee and I would rest on hot afternoons in a little grove of aspens that had somehow gotten started in this desolate place that everyone wanted to cover in garbage. We know now that these aspens originated from seeds caught on a patch of moss, and the whole island of shade began to grow from there. The trees brought birds and the birds brought berries—raspberries, strawberries, blueberries—which now blossom around us. The center of the grove was cool and shady, and the leaf litter from the aspens had started to build up a thin layer of soil over the tailings. Sheltered from the harsh conditions of the mine, a few maple seedlings, migrants from the surrounding forest, were holding their own. Brushing aside the leaf litter, we uncovered the remnants of *Polytrichum*, the first plants to begin healing the land, making it possible for others to follow. In the deepening shade, they'll soon be replaced, having done their work. This island of trees is the legacy of mosses pioneering on the tailings.

In the Forest of the Waterbear

"Mysterious and little-known organisms live within reach of where you sit. Splendor awaits in minute proportions."

E. O. Wilson

The rain forest beckons to botanists like Mecca to the faithful. For years I dreamed of a journey to the cradle of plant civilization, the verdant Holy Grail. When the time came for my pilgrimage, my head was filled with wild anticipation of outlandish beings and greenery beyond imagination. The Amazon called and I followed, by plane to the waiting jeep, to the dugout down the muddy river, and at last on foot, into the dripping forest.

The interior of the rain forest is overwhelming in its complexity. There is not a bare surface anywhere. Branches are hung with curtains of mosses and sprays of orchids dangle among them. Tree trunks are filmed over with algae, studded by giant ferns, and wound about with vines. Ants travel in convoys across the ground and up the trees, and metallic beetles glint in sun-flecks on the forest floor. The forest itself is richly textured; stems embossed with every manner of protuberance, leaves ornamented by spines and pleats, scales and fringes. Long shafts of sunlight cut through the dark canopy and catch the flash of iridescent butterfly wings before diffusing in the vegetation below.

While the jungle was overwhelmingly exotic, I was haunted by a sense of having seen it all before. There was something strangely familiar about the quality of the light; leafy, wet, and saturated with green. The pervasive shadows and movements at the periphery of vision brought forth a familiar sense of possibility, and the desire to part the undergrowth and go wandering. It reminded me of walking through a moss.

This is entirely possible with a good stereomicroscope, which lets you go wandering at will through a living moss turf, like bushwhacking through the jungle. Tiny needle in hand, like a machete to make a path, or a walking stick to part the palm fronds, I've spent hours lost in looking, threading my way between the stems, bending low beneath a branch and turning over leaves to see what's underneath. The stereomicroscope provides a way into the forest of a moss clump, in three dimensions. I can zoom in for a closer look, or step back to see the panorama.

I'm struck by the parallels between the moss microcosm and the rain forest. The similarities are more than visual. While the height of the moss mat is approximately three thousand times smaller than that of the rain forest, they nonetheless exhibit the same kind of structure, the same kind of function. Like those in the rain forest, the animals of the moss forest are interconnected in complex food webs: herbivores, carnivores, and predators. The ecosystem rules of energy flow and nutrient cycling, competition, and mutualism still apply here. The patterns clearly transcend the vast difference in scale.

Accustomed to the benign nature of northern forests, I had to continually remind myself not to push through the jungle vegetation without looking first. Grabbing a branch might mean a sting from a conga ant that would lay you low for the next 24 hours. Stepping over a log without looking might yield an encounter with a fer de lance snake that could lay you low forever. Our Quechua guides taught us the three things one must take into the forest for safety: the eyes, the ears, and the machete. Most of the plants were surprisingly well armed. Toothed leaves, spiny stems, and prickly bark were the norm and my hands had enough scratches and punctures to make me attentive every time I walked through the forest. Dwarfed by the green, vulnerable, I felt something in common with the tiny creatures in a moss mat. I could imagine how a soft-bodied larva feels as it twists and turns through the dense stems of a moss mat, where the leaves are sharp pointed and edged with teeth.

My Ecuadorian colleagues brought us to a canopy observation platform in the ecological reserve. We climbed one at a time up a winding set of narrow stairs built around a gigantic *Ceiba* tree that rose up through the canopy and punched a hole through to the sky. The canopy world is usually accessible only to the birds and bats, and now a few lucky scientists. With every spiraling turn around the tree, we moved through the complex layers that make up the forest.

The rain forest canopy supports a lush flora of epiphytes, plants living on the trunks and branches in the full tropical sun, getting their water from the rain and their nutrients from the air. Ferns and orchids carpet the branches and lianas twine around the trunks and bind them together in a tangle of vines. Ahead of me, just beyond arm's reach, is a garden of bromeliads, with waxy red leaves looking like flowers. The leaves overlap one another to create pockets that collect the rain, which comes every afternoon, predictably at 2:00. There are species of mosquitoes, and even frogs, that complete their entire life cycle in these bromeliad tanks, high above the forest floor. Far from the soil, mosses are the foundation for most of these epiphytes, forming a deep cushion all along the tree branches.

Mosses are not only epiphytic on other plants; they support epiphytes of their own. The interior of a moss clump can be heavily colonized by algae, making it look like a moss-draped rain forest in miniature. Golden disks of single-celled algae rest among the moss leaves. Threads of tiny liverworts coil around the stems like vines on a tree trunk and competing mosses may engulf a stem like a strangler fig. Clinging to the rhizoids of the moss are colorful spores and pollen grains, evoking the pattern of pastel orchids. The moss forest even has its equivalent of bromeliad tanks. The water-filled pocket in a moss leaf can support unique species of rotifers, invertebrates that know no other home but the tiny pool among the moss leaves.

A hallmark of the tropical rain forest is the intense vertical stratification from canopy top to the surface of the soil. The flora and fauna are adapted to the gradient of sunlight, intense at the surface and diminishing as it passes through layers of the forest to the deep shade of the forest floor. Fruit-eating bats cruise the top of the canopy, while bird-eating tarantulas hide in dim light among the buttress roots. The moss forest is stratified in a similar fashion. Some insects frequent the dry open top of the clump, while others like springtails burrow deep in the damp rhizoids at the bottom.

While one is walking in the rain forest, there is a steady pitter-patter, not of raindrops, but of bits of debris falling from the canopy. Old leaves, bugs, and spent petals are constantly drifting downward, enriching the soil and recycling nutrients from the producers at the forest top to the decomposers at the bottom. We were repeatedly startled when half-eaten fruit would come plummeting down from above, the remnants of a parrot's meal. Fruits or nuts falling from the high canopy can really pack a wallop on a bare head. Our guide displayed his egg-shaped bruise. If you could walk at the bottom of a moss colony, there would be the same steady rain of particles through the layers of leaves. The moss turf traps windblown soil, and leaf fragments, dead bugs, and spores, which collect at the base of the moss, gradually building up soil where none had been before. Decaying organic matter hosts filaments of fungi, which are greedily fed upon by springtails. It is this accumulation of decaying debris that provides an anchor for rooted plants, akin to orchids in the rain forest or ferns taking hold on a mossy rock.

A Berlese funnel is the tool typically used to study the nearly invisible fauna of microcommunities such as moss. Soil, rotting wood, or a clump of moss is put into a large aluminum funnel, fitted with a screen. A barrage of high-intensity lamps is placed over the top of the funnel for several days. Slowly, the heat begins to dry out the moss or other material. Fleeing the light and seeking out the remaining moisture, all the invertebrates move downward toward the tip of the funnel where they fall to their deaths, collected in a jar of formaldehyde.

The collection from a Berlese funnel might typically produce the following results. One gram of moss from the forest floor, a piece about the size of a muffin, would harbor 150,000 protozoa, 132,000 tardigrades, 3,000 springtails, 800 rotifers, 500 nematodes, 400 mites, and 200 fly larvae. These numbers tell us something about the astounding quantity of life in a handful of moss.

But the numbers themselves miss the point. Such lists remind me of the inconsequential facts tossed off by a tour guide, the number of steps to the top of the Washington monument or the number of granite blocks used to construct it, when what I really want to know about is the view from the top and the jokes told by the stonemasons. Berlese funnels yield a good inventory of the biota, I suppose, but I'd rather go walking through a moss clump and see the thousands of creatures living out their lives than count their bodies in a jar.

Invertebrates are attracted to forests of moss for the same reasons that rain forests shelter such diversity of wildlife. They offer a favorable microclimate, shelter, food, nutrients, and a complex internal structure, which creates a great diversity of habitats. And like the rain forest, a moss forest is a hotspot for evolution. Mosses were the first plants to colonize the land, and paved the way for the creatures that followed. Many entomologists believe that the early stages of insect evolution took place in the mats of moss. The moist protection offered by mosses created a transitional environment between primitive aquatic life and more advanced terrestrial organisms. Today, many advanced insects still rely on moss mats for nurture of their eggs and larvae. Craneflies hover around mossy cliffs, waiting to deposit their eggs in the wet leaves. The cranefly mothers are quite selective in choosing a nursery for their offspring. They avoid mosses with sharp leaves and densely packed stems which would make life difficult for the tunneling larvae.

Each morning in the jungle, we'd awake to the sound of parrots, squawking through the canopy, vivid as a kindergarten paintbox. Long tail feathers streaming behind, the red of a scarlet macaw is startling against the green leaves. The forest of mosses has its own brilliant spots of color moving through the branches. Here the red belongs to Oribatid mites. Round and shiny, the mites remind me of eight-legged bowling balls scurrying over the foliage. When my probing disturbs them, they simply veer off in another direction, and I follow them along as they forage for spores, algae, and protozoa. Some of the mites are predaceous on other invertebrates, and some eat the moss foliage.

Amazon nights come quickly when the sun drops below the equator, without the interlude of twilight. At the fall of darkness, we'd return to the bamboo platform, which was our camp. The shelter was raised on stilts and we climbed to the platform up an inclined log with steps cut into it. Before we blew out the candle for the night, the log steps were pulled up to discourage unwanted visitors. Falling asleep was a challenge, despite the exhausting days of hiking in the tropical heat. The night was alive with sounds: frogs bellowing, toads trilling, insects buzzing, and one night, the yowling of a panther.

Predators lurk in the moss forest, too. Pseudoscorpions conceal themselves among the dead leaves and dart out on rippling rows of legs to sting their prey. Carabid beetles, hard shelled and shining, patrol the moss turf with their enormous pincers and take small invertebrates wherever they find them. Predaceous larvae lie like snakes in the branches.

The intensity of predation in the rain forest has led to many adaptations for camouflage and mimicry. There are moths that resemble dead leaves, snakes that mimic branches, and caterpillars disguised as bird droppings. So, too, in the moss forest, there are creatures disguised as bryophytes. In New Guinea, moss weevils carry about tiny gardens of moss on their backs, growing in special cavities on their shells. The larvae of some craneflies are a mossy green color, marked with dark lines to hide among the leaves. They move sluggishly through the moss mat, further concealing their presence by lethargy. This same approach to predator avoidance is used by the tree sloth in the jungle, coated with algae and moving so slowly that it becomes nearly invisible in the canopy.

The dense foliage works to the advantage of predator and prey that do not wish to be seen. But this same concealing profusion can be a liability when the intention is sexual display. Life in the jungle depends on the imperative for reproduction, to somehow find the right mate in a habitat already saturated with life. Birds resolve this dilemma by adopting gaudy plumage and loud calls, which penetrate the forest, advertising their availability. Likewise, every plant seems to be locked in competition to be noticed, to seduce potential pollinators to carry pollen to the next blossom. The fate of many plant species lies in the hands of complex interactions with pollinators, butterflies, bees, bats, and hummingbirds. Hummingbirds abound in the canopy, their iridescence dazzling in the sunlight. They move like dragonflies, zipping so quickly from flower to flower that you can hardly get a look at them. My best chance for close observation came when a jewel-like hummingbird hovered near the red baseball cap of a fellow hiker. He could hear the hum and feel the breeze from the beating wings and we all silently begged him not to move, as the bird delicately probed the strange new Red Sox blossom that had appeared in its territory.

Mosses experience the same pressure for cross-fertilization, but lack flowers or any kind of showy display to attract insects as co-conspirators in fertilization. Mosses rely on the movement of water to carry sperm, and it is an inefficient process since the sperm can rarely travel more than a few centimeters. It appears, however, that the invertebrate community that inhabits mosses has the potential to carry sperm a bit farther afield. As they crawl through the moss, mites, springtails, and other arthropods passing by a male can become smeared with mucilage containing moss sperm. The sperm may then be carried on the bodies of the invertebrates and rinsed off into water droplets elsewhere in the moss, where they can swim to the waiting females. Invertebrates are unwitting but vital partners in the continuity of the moss forest, like the hummingbird with pollen accidentally brushed on its forehead.

The bright colors of tropical flowers are repeated in their fruits. The most common color for fruits in the canopy is red, since that is most visible to birds and to monkeys, the most important dispersers of seeds. Dispersal in mosses is typically by wind, although one species, *Splachnum*, has evolved a brightly colored sporophyte and strong odors to attract dung flies, which then carry its spores. Birds, mammals, and especially ants often feed upon the protein-rich sporophytes. I've watched a sparrow systematically harvest a crop of sporophytes from haircap moss, neatly clipping off the capsules in its beak and trailing a cloud of spores. Ants are no doubt good dispersers of moss, as they carry the open capsules on their backs, sprinkling spores all along the route to their nest.

Development and population pressure in the rain forest have triggered a sharp decline in wildlife populations. Therefore, our guides were very excited to pick up the tracks of a mother tapir and her baby in the mud. We woke before dawn the next day to follow their trail along the river in hopes of seeing them. In the misty stillness of the early morning, we wove our way through the palms of the riverine forest, listening intently. The tapirs had vanished, but quiet walking in the woods never disappoints. We heard a troop of howler monkeys waking and watched them move through the branches overhead, perfectly adapted for life in the treetops.

Moving quietly through the microscopic forest, peering between branches and following a glimpse of movement, I'm on the trail of a tardigrade. If I had to choose one animal whose life is most closely tied to the life of mosses, it would be the waterbear, or tardigrade. Like the panda bear, which is totally reliant on the bamboo forest, the waterbear's life is inextricable from the moss in which it lives. Nosing through the foliage, trundling along on eight stumpy legs, the waterbear bears a remarkable likeness to a tiny polar bear. Low slung, with a round head, its body translucent and pearly white, the waterbear clings by long black claws to the moss stems. Instead of a jaw full of teeth, a waterbear has sucking mouthparts. It feeds by piercing a moss cell with a stylet like a hypodermic needle, and sucking out the contents of the cell. Other types of tardigrades graze on algae and bacteria, the epiphytes of moss leaves. A few are even predaceous, using the stylet on other invertebrates, sucking out their cells.

As their name suggests, waterbears are reliant on the abundant moisture held in the interstitial spaces of a moss clump. They cross between plants on fragile bridges of water, spanning the capillary spaces in the moss. A typical place I go to look for them is in a moss with deeply concave leaves. The tiny pool of water held in a spoon-shaped leaf is the perfect resting place for a waterbear, as plump and gelatinous as a candy Gummi Bear. The moisture in a moss mat is as vital to the moss as it is to the waterbear. But, since mosses are nonvascular, their water content fluctuates with the amount of water in the environment. The moss leaves shrivel and contort as water evaporates, leaving them crisp and dry. The waterbears, too, simply shrink when desiccated to as little as one-eighth of their size, forming barrel–shaped miniatures of themselves called tuns. Metabolism is reduced to near zero and the tun can survive in this state for years. The tuns blow around in the dry winds like specks of dust, landing on new clumps of moss and dispersing farther than their short waterbear legs could ever carry them.

Neither the moss nor the waterbear is damaged in the process of desiccation. In this state of suspended animation, they are invulnerable to extremes of temperature or other environmental stresses. The moment that fresh water becomes available, as dew or a welcome rain shower, the waterbear and the moss soak up the water and swell back to their normal size and shape. Within twenty minutes, the moss and the waterbear, in perfect synchrony, resume their normal activities.

Rotifers, or "wheel animalcules" as they were first called, share this same remarkable ability to withstand drying. When moist, rotifers inhabit the water-filled spaces of a moss, like guppies in a multitude of tiny aquaria. They can be easily seen feeding there as their rotating cilia draw in food particles on a current generated by the spinning "wheel" of their mouths.

Within the moss microcosm, evolution has produced a shared adaptation to the inevitable fluctuation in moisture. Just as the evolution of birds is tied to the evolution of the trees they live among, the lives of waterbears and rotifers has been shaped by adaptations of the mosses.

All three—mosses, waterbears, and rotifers—figured prominently in a nineteenth-century debate about revivification and the very nature of life. The behavior of these three blurs the distinction at the edge between life and death. All signs of life are extinguished when they are dry: no movement, no gas exchange, no metabolism. All enter a state known as anabiosis, or lack of life. And yet, as soon as water is returned, life suddenly is renewed. Their apparent death, followed by resuscitation, suggested that life might be stopped and then re-started. Waterbears were the subject of intense experimentation to test the limits of their endurance. In the dry state, they were subject to conditions that would kill any known organism: boiling, being held in a vacuum only 0.008 degrees above absolute zero. But, without fail, they tolerated these abuses and were revivified with a drop of water. The addition of water unlocks the chemistry of life by a mechanism that is still largely unknown, but utilized by mosses and waterbears every day.

After 350 years of lively debate and experimentation, it is generally agreed that life does not cease in anabiotic organisms, but continues at a barely perceptible rate. Sophisticated technologies are required to document the infinitesimal rate of metabolism, which permits life to be suspended indefinitely. The process that allows these beings to hover at the boundary between life and death is still a profound mystery that is continually played out in the mosses beneath our feet.

It took a flight across the equator, a perilous crossing of the Andes and three days down the river in a dugout canoe to bring me to the heart of the rain forest. But at home I don't have to go that far to find a shadowy forest full of exotic beings that I've never seen before. In a five-minute walk down the path of my garden I can have a handful of moss, and a five-minute walk back to the microscope brings me to the lush interior of the moss forest. There is no word but awe for the biological excess of that place, the profusion of life, vivid and complex beyond our grasp. At every turn of a leaf, there are mysteries. There are life forms here that occur nowhere else on the planet and intricate relationships evolved over eons. You might take care not to step on them.

Kickapoo

I finally got around to refinishing the bottom of my canoe. After the duct tape wore off. Ahh, duct tape, the great enabler of the procrastinator. I peel it off, layer after layer, where I'd slapped it on after a collision with a rock on the Oswegatchie, and where the stern bumped down hard on a ledge of the New River. Inspecting the various cracks and chips is like taking inventory of great canoe trips. Here's a souvenir of the rapids on the Flambeau and here the gravel beds of the Raquette. Along the gunnel there is a smudge of red paint, running for six inches or so along the sky blue fiberglass. I puzzle over that one for a moment and then I remember the Kickapoo and the summer I spent immersed.

The Kickapoo River runs through southwestern Wisconsin in a region known as the Driftless Area. The glaciers which covered the upper Midwest skipped this one little corner of Wisconsin, leaving a landscape of steep cliffs and sandstone canyons. I discovered the stream with a fellow graduate student as she surveyed the area for rare lichens. We paddled down the river, stopping at cliffs and outcrops to scan the species. All along the river I was struck by the distinctive pattern on the cliffs. The upper reaches of the cliff were spattered with lichens, but at the foot of the sheer wall were horizontal bands of moss in different shades of green, rising from the water. I was looking to find a thesis question and this one found me. What was the source of the vertical stratification that striped the cliff?

I had some ideas, of course. I'd climbed too many mountains not to notice the changes in vegetation with elevation. Elevational zonation usually results from temperature gradients and it gets cooler the higher you go. I imagined that there would be some kind of environmental gradient that changed as the cliff rose from the water, and the moss pattern would follow.

The next week I went back to the Kickapoo by myself, ready to look more closely at the banded cliffs. I put my canoe in at the bridge and paddled upstream. The current was swifter than it looked and I had to paddle hard. I maneuvered alongside the rock face, but there was nowhere to moor the canoe. Every time I stopped paddling to look at the mosses I'd be pulled downstream. I could hang on with my fingers wedged in a crack, just long enough to snatch a clump of moss, and then I'd drift away again. Any kind of systematic study was clearly going to require a different approach.

I beached the canoe on the opposite bank and decided to see if I could wade over to the cliff. The bottom was sandy and the river only knee deep. The cool water, swirling around my legs, felt wonderful on a hot day. This was starting to feel like the perfect research site. I waded over within arm's reach of the cliff. Suddenly, the bottom dropped away. The current had undercut the cliff and I found myself chest deep and clinging to the rock. But what a great face-to-face view of the mosses.

Right next to the water, extending upward for a foot or so, was a dark band of *Fissidens osmundoides*. *Fissidens* is a small moss. Each shoot is only 8 mm high, but it is tough and wiry. *Fissidens'* form is very distinctive. The whole plant is flat, like an upright feather. Each leaf has a smooth thin blade, atop which sits a second flap of leaf, like a flat pocket on a shirtfront. This envelope of leaf seems to function in holding water. All crowded together, the shoots make a rough-textured turf. *Fissidens* has well-developed rhizoids, root-like filaments that attach firmly to the grainy sandstone. At the waterline *Fissidens* formed a virtual monoculture. I saw hardly any other species, save a snail or two hanging on for dear life.

Fissidens

About a foot above water, the *Fissidens* disappeared and was replaced by assorted clumps of other mosses. Silky tufts of *Gymnostomum aeruginosum*, mounds of *Bryum* and glistening mats of *Mnium*, all are arrayed in a patchwork of different greens amidst empty patches of tawny sandstone.

Higher still, just at the limit of what I could reach from my underwater perch, began a dense mat of *Conocephalum conicum*, a thallose liverwort. Liverworts are primitive relatives of mosses. They get their unappealing name from the botany of the Middle Ages. "Wort" is the old Anglo-Saxon word for plant. The medieval Doctrine of Signatures proposed that all plants had some use to humans and would give us a sign to reveal that use: any resemblance between a human organ and the plant would suggest it as a remedy. The leaves of liverworts are generally three-lobed, as is the human liver. There is no evidence that liverworts made effective cures, but the name has persisted for seven centuries. In the case of *Conocephalum* a better name might be snakewort, for it bears a close resemblance to the scaly skin of a green adder. This plant has no distinct leaves, just a sinuous, flattened thallus ending in three round lobes like the triangular head of a viper. Its surface is divided into tiny diamond-shaped polygons, contributing to its reptilian appearance. Closely appressed to the surface, it snakes its way over rock or soil, held loosely in place by a line of scraggly rhizoids on its underside. Brilliant green, exotic, *Conocephalum* completely covers the cliff at this height, making a striking contrast to the darker mosses below.

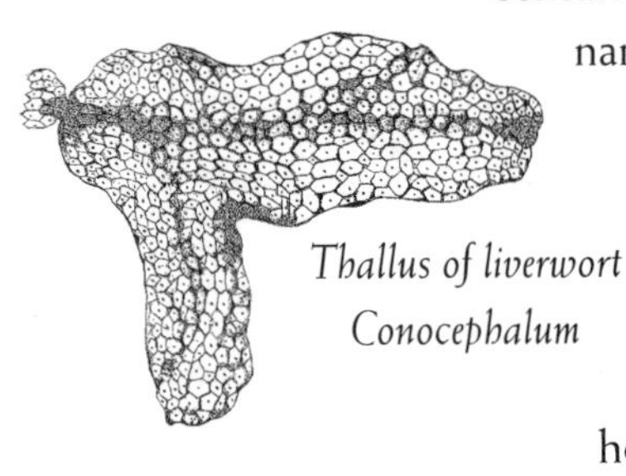

Thallus of liverwort Conocephalum

I was captivated by these plants and their layered distribution on the cliff. The fact that I could paddle to my research site cinched my choice of thesis topic. The only problem was logistics. How could I make all the detailed measurements I needed while chest deep in the river? Over the next few weeks I tried lots of things. I tried anchoring the canoe and leaning out toward the cliff. The number of dropped pencils and rulers was disheartening, as was the constant threat of capsizing. I tied little Styrofoam floats to all my equipment, but the current just carried them away, bobbing merrily downstream before I could grab them. So I tethered all my gear to the thwarts of the canoe and you can imagine the resulting tangle of camera straps, data books, and light meters. Eventually, I abandoned ship and simply planted my feet on the river bottom. I devised a kind of floating laboratory with the canoe anchored beside the cliff and me standing in the river where I could reach both rocks and canoe. Data books were impossible to manage. I kept dropping them in. So I collected my measurements using a tape recorder. The machine sat securely duct-taped to the seat of the canoe and the microphone was looped around my neck. I could then have both hands free to position my sampling grids and collect specimens, and still have a free leg to snare the canoe rope when it began to drift. I felt like the one-man band of the Kickapoo. It must have made quite a picture as I was talking to myself, immersed in the river, and singing out the locations and abundance of the mosses: *Conocephalum* 35, *Fissidens* 24, *Gymnostomum* 6. I marked all the plots with dabs of red paint, which still decorates my canoe.

In the evenings I'd transcribe the tapes, converting my recorded litany to real data. I wish I'd kept some of those tapes, just for entertainment value. In between the hours of droned numbers were bursts of inspired cursing as the canoe started to drift away, tightening the microphone around my neck. I recorded any number of squeals and frantic splashes when something nibbled at my legs. I even had tape of an entire conversation with passing canoeists who handed me a cold Leinenkugels Ale as they floated by.

The vertical stratification of species was very clear with *Fissidens* at the bottom, *Conocephalum* at the top, and a variety of others sandwiched in between. But my hypothesis about the cause of the pattern was not supported. There were no significant differences in light, temperature, humidity, or rock type along the face of the cliff. The pattern had to be caused by something else. Standing in the river day after day, I was becoming vertically stratified myself—shrivelled toes at the bottom, sunburned nose at the top, and muddy in between.

Oftentimes, an abrupt pattern in nature is caused by an interaction between species, such as territorial defense or one tree species shading out another. The pattern I was observing might well be the result of some competitive "line in the sand" between *Conocephalum* and *Fissidens*. I gave the two species a chance to tell me about their relationship, by growing them side by side in the greenhouse. Alone, *Fissidens* did fine. *Conocephalum*, likewise. But when they were grown together there was clear evidence of a power struggle, which was consistently lost by *Fissidens*. Time after time, *Conocephalum* extended its snaky thallus over the top of diminutive *Fissidens*, completely engulfing it. Their separation on the cliff became clearer. *Fissidens* had to keep away from the liverwort in order to survive. But, if competition was so important, why didn't *Conocephalum* grow all the way to the waterline and simply obliterate the other species?

One day in late summer I noticed a wad of grass snagged on a branch high above my head—a high-water mark. Clearly the river was not always at wading depth. Perhaps the vertical stratification was due to differences in how the species tolerated flooding. I collected clumps of each species and submerged them in pans of water for various times: 12, 24, 48 hours. The *Fissidens* remained perfectly healthy even after three days, as did *Gymnostomum*. But after only 24 hours the *Conocephalum* was black and slimy. So here was a piece of the pattern. *Conocephalum* must be confined to the higher levels of the cliff by its inability to withstand flooding.

I wondered how often floods like the one I'd simulated actually happened. Could it be often enough to create a barrier for *Conocephalum*'s expansionist tendencies? As luck would have it, the Army Corps of Engineers was wondering the same thing, albeit for a different reason. They were considering constructing a flood-control dam on the river and had installed a gaging station at the bridge below my cliffs. They had amassed five years of daily measurements of water levels on the Kickapoo. I could use their data to calculate the frequency with which any point on the cliff had been underwater. I could also call in to the automated phone number to learn the current water level at the bridge. I've not been much of a cheerleader for the Corps, given their propensity for spoiling rivers, but these data were invaluable.

All winter long I analyzed the data to match them to the distribution of mosses on the cliff. Not surprisingly, the gaging station data matched the elevational zonation of the bryophytes very well. The water level was most frequently lapping at the base of the cliff where *Fissidens* dominated the vegetation. It was tolerant of flooding and its wiry streamlined stems allowed it to withstand the frequent company of the current. Flood frequency declined with rising elevation on the cliff. The zone dominated by loosely attached *Conocephalum* was inundated very rarely. High above the water, *Conocephalum* could safely spread its snaking thallus over the rock in an uninterrupted blanket of green. One species dominated where flood frequency was high. Another species dominated where disturbance was low. But what about in the middle? Here was a tremendous variety of species, as well as patches of open rock as bare as a billboard advertising "space available." In the zone of intermediate flood frequency no one species dominated and diversity was high. As many as ten other species were sandwiched here between the two superpowers.

At the same time as I was wading the Kickapoo, another scientist, Robert Paine, was exploring a different gradient of disturbance frequency, wave action on the rocky intertidal zone of the Washington coast. He was looking at communities of algae, mussels, and barnacles, which may seem to have little in common with mosses. And yet both are sessile, attached to rock and engaged in a competition for space. He observed an intriguing pattern—-few species lived where the wave action was constant and fewer still lived on rocks which were virtually undisturbed. But in between, where disturbance was intermediate in frequency, species diversity was extremely high.

The rocky coast and the Kickapoo cliffs helped to generate what has become known as the Intermediate Disturbance Hypothesis, that diversity of species is highest when disturbance occurs at an interval between the extremes. Ecologists have shown that in the complete absence of disturbance, superior competitors like *Conocephalum* can slowly encroach upon other species and eliminate them by competitive dominance. Where disturbance is very frequent, only the very hardiest species can survive the tumult. But in between, at intermediate frequency, there seems to be a balance that permits a great variety of species to flourish. Disturbance is just frequent enough to prevent competitive dominance and yet stable periods are long enough for successional species to become established. Diversity is maximized when there are many kinds of patches of all different ages.

The Intermediate Disturbance Hypothesis has been verified in a host of other ecosystems: prairies, rivers, coral reefs, and forests. The pattern it reveals is at the core of the Forest Service's policy on fire. Fire suppression with Smokey Bear's vigilance produced a disturbance frequency which was too low and the forests became a monocultural tinderbox. Too high a fire frequency left only a few scrubby species. But like Goldilocks and the Three Bears (one must have been Smokey himself), there is a fire frequency which is "just right," and here diversity abounds. Creation of a mosaic of patches by mid-frequency burning creates wildlife habitat and maintains forest health, while fire suppression does not.

When the ice went out on the Kickapoo the next spring I called the gaging station and an electronic voice informed me that the river was in flood. So I jumped in my car and drove down to see what the mosses looked like now. The river was chocolate brown with dissolved farmland. Logs and old fenceposts were pushed along in the torrent, bumping against the cliff. My red paint markers were nowhere to be seen. By the next morning the waters had receded as quickly as they had come and the aftermath was revealed. The *Fissidens* had emerged unscathed. The mid-level mosses were sodden with mud and battered by the logs and the pull of the water. A few more bare patches had been made. The *Conocephalum* had not been submerged long enough to die, but it was torn away in great swaths, hanging from the cliff like ripped wallpaper. Its flat loose form had made it particularly vulnerable to the pull of water, while *Fissidens* was unaffected. The open patches created by the removal of *Conocephalum* made temporary habitats for a new generation of mosses which would persist there until *Conocephalum* gathered its strength and returned. These are the species which are not able to compete with *Conocephalum*, nor to withstand the frequent flooding. They are fugitives between two forces, living in the crossfire between competition and the force of the river.

I like to think of the satisfying coherence in that pattern. Mosses, mussels, forests, and prairies all seem to be governed by the same principle. The apparent destruction of a disturbance is in fact an act of renewal, provided the balance is right. The Kickapoo mosses had a piece in telling that story. Sandpaper in hand, I look at the splotch of red paint on this old blue canoe and decide to let it be.

Choices

My neighbor, Paulie and I communicate mostly by shouting. I'll be outside unpacking the car and she'll stick her head out of the barn and yell across the road, "How was your trip? Big rain while you were gone, the squash in the garden are going crazy—help yourself." Her head pops back into the barn before I can answer. She takes a dim view of my gallivanting around, but keeps a good eye on the place while I'm gone. While I'm out stacking firewood or planting beans, I'll catch sight of her blaze orange cap and call across the road to her with news of a downed fenceline I discovered up by the pond. Our shouts carry the shorthand affection we have for each other. Over the years it's been a telegraph from my side of the road to hers, carrying messages of kids growing up, parents growing old, breakdowns of the manure spreader, and news of the killdeer nesting in the pasture. On 9/11 I ran from my TV to the barn where we hugged and cried for a short moment until the feed truck arrived and brought us back to the immediate need of calves to be fed.

My old house and her old barn, in the little town of Fabius, New York, were once part of the same farm, starting way back in 1823. They share the shade of the same big maples and are watered by a common spring. We've brought them back from the brink of decay together, so it's fitting that we, too, are friends. Sometimes, when the weather is nice, we stand with arms folded in the middle of the road to talk, shooing barn cats out of the road and holding up traffic, which consists of the occasional haywagon or the milk truck. Our dirty work gloves are pulled off as we soak up the sun and the talk and are pulled back on again as we turn away. On the rare occasions when we do talk on the phone she forgets she's not hollering from the barn and I have to hold the phone a foot from my ear.

As observant neighbors we know a lot about each other. She just shakes her head and laughs over my field seasons spent earnestly investigating the reproductive choices of mosses. All the while she and her husband Ed are milking 86 head, raising corn, shearing sheep, and building a heifer barn. Just this morning, we met down at my mailbox and had a moment to talk while she was waiting for the AI man. "Artificial Intelligence?" I asked with a raised eyebrow. This cracked her up, one more sign of the detached ignorance of her neighbor, the professor. The white panel truck splashed over the potholes to the barn, a picture of a bull on the side. "Artificial Insemination," she shouted over her shoulder as we walked back into our worlds on opposite sides of the road. "Your mosses may have reproductive choices, but my cows sure don't."

Mosses do exhibit the entire range of reproductive behaviors from uninhibited sexual frenzy to puritanical abstention. There are sexually active species churning out millions of offspring at a time and celibate species in which sexual reproduction has never been observed. Transexuality is not unheard of; some species alter their gender quite freely.

Plant ecologists measure a plant's enthusiasm for sexual reproduction with an index known as reproductive effort. This measure is simply the proportion of the plant's total body weight which is dedicated to sexual reproduction. For example, our maple tree allocates much more of its energy to production of wood than to its tiny flowers and helicopter seeds that twirl to the ground on the breeze. In contrast, the dandelion in the pasture has a very high reproductive effort, with much of the plant's mass tied up in yellow flowers, followed by drifts of fluffy seeds.

The energy allocated to reproduction can be spent in a variety of ways. The same number of calories could be used to make a few large offspring that the parents invest in heavily. Alternatively, some are more profligate, spending their energy on a large number of tiny, poorly provisioned offspring. Paulie has strong opinions on those who have children that they don't adequately support. One of the barn cats, a long-haired beauty named Blue, seems to take the attitude that kittens are a disposable commodity. She has litter after litter, but can't be bothered to nurse them and leaves them to fend for themselves. Mosses like *Ceratodon* take the same approach. On a patch of disturbed ground along the cow track to the barn, the leaves of *Ceratodon purpureus* are barely visible under the dense swath of sporophytes it produces all year long. But each spore is so small and poorly provisioned that, like Blue's kittens, it has a vanishingly small chance of surviving. Fortunately, there is among the barn cats a paragon of good mothering, Oscar. She's the old lady of the haymow, and carefully tends her single litter, and willingly adopts Blue's orphans as her own. For this, Oscar earns a place at the milk dish at milking time.

Paulie would approve of a moss like *Anomodon*, growing on the shaded rock wall behind the barn. This species delays its spore production until later in life, preferring to allocate its resources to growth and survival, rather than unfettered reproduction.

The two strategies of high and low reproductive effort are usually associated with a particular environment. In an unstable, disturbed habitat, evolution will tend to favor those species that produce many small highly dispersable offspring. The unpredictable nature of the habitat, like the *Ceratodon* near a cow path, means that the adults have a high risk of dying by disturbance, and so it is advantageous to reproduce quickly and send your progeny off to greener pastures. The destination of those wind-blown spores is unknown, but is likely quite different from the path edge of the parents. Sexual reproduction also conveys a strong advantage by mixing up the parents' genes into new combinations. Every spore is like a lottery ticket. Some will be good combinations, some will be bad, but the gamble pays off with millions of offspring spread randomly over the landscape. One will surely find a patch of ground where its novel genetic formula will bring it success. Sexual reproduction creates variety, a distinct advantage in an unpredictable world. However, sexual reproduction also incurs some costs. In creation of egg and sperm, only half of the parents' successful genes are passed to the offspring, and those genes are shuffled in the lottery of sexual reproduction.

In her muddy boots and manure-spattered jacket, Paulie doesn't fit the white-coat image of genetic engineering, and yet she is working at the forefront of its application. A Cornell grad, she has bred an award-winning herd of Holsteins with impeccable genetic pedigrees. Rather than lose this hard-won genetic advantage by mating her best cows with any old bull, she is using artificial insemination and then transferring the identical embryos to surrogate mothers in the herd. In this way, she will develop a herd with little variability, perpetuating the successful genotypes that would have been scrambled by ordinary sexual reproduction. Such cloning is a recent development in dairy production, but mosses have been doing it since the Devonian era.

Reproductive strategies that limit variation and preserve the parents' favorable gene combinations are commonplace among the mosses. The rock wall behind the barn has been undisturbed since the first farm owners built it 179 years ago. In such a steady, predictable habitat a steady, predictable way of life is most successful. The *Anomodon* mat that lives there has had nearly two centuries to prove that it bears a genetic makeup well suited to that particular spot. Energy devoted to frequent sexual reproduction would essentially be wasted here, by producing wind-blown spores of potentially unfit genotypes, which would simply be lost on the wind. In a stable, favorable environment, it is better to invest that energy in growth and clonal expansion of the existing long-lived moss, preserving the tried and true genotype, like pedigreed cows.

Natural selection is constantly acting upon the pool of individuals that make up a population, and only the most fit survive. Burying generations of barn cats who never learned to cross the road, or stillborn calves, clearly reveals the hand of natural selection. On such occasions, Paulie brushes off the loss with a practiced line. "If you're going to have livestock, you're going to have deadstock." Despite her bluster, Paulie's menagerie tells a different story. Not all of her animals are the cream of the crop. One stall is home for an old cow, blind now for many years. Her name is Helen. She's a good old girl and with the time-honored nose-to-tail guidance system she stills goes out to pasture with the others. And then there's Cornellie, the orphaned lamb whom Paulie brought home in diapers to sleep by the woodstove until it was big enough to survive. But, in nature, there is rarely a Paulie who spares the unfit from the scythe of natural selection. So I've been looking at the reproductive choices made by mosses in light of natural selection. Which choices result in survival and which are steps toward extinction?

Chance and our choices have brought Paulie and I together, converging for some reason on this old hill farm. Something about the way the house nestles in the hill sheltered from the wind, or the way the morning sun pours over the meadow. She fled the expectations of

Boston family and chose the intense flavor of farming over a career as an animal physiologist. I flew here like a homing pigeon after a sad divorce with the fervor to start again, on my own terms. Our dreams have found a home here. Paulie recreates her self-sufficiency every single day and revels in the company of animals. And here my microscopes can share the table with blackberry pies.

Up in the hemlock swamp at the top of our pasture, the woods are fenced from grazing. Paulie is mowing hay in the adjacent field, the tractor rumbling along. I wave to her as I duck under the barbed wire and into the woods. A few steps into the trees and a hush descends with the green filtered light. The hemlock timbers which built my house and Paulie's barn were cut here generations ago. The old logs and decaying stumps are covered with one of my very favorite mosses, *Tetraphis pellucida*. I know of no moss more charged with well-being than *Tetraphis*. Its young leaves are luminous as dewdrops and swollen with water. The species epithet "*pellucida*" reflects this watery quality of transparency. Its sturdy little shoots are clean and simple and stand upright in a hopeful sort of way. Each stem is no more than a centimeter tall with a dozen or so spoon-shaped leaves arranged like an open spiral staircase ascending the stem.

In contrast to most mosses, which have adopted a particular lifestyle and stuck with it, *Tetraphis* is remarkable for its flexibility in making reproductive choices, sexual and otherwise. *Tetraphis* is unique in having specialized means of both sexual and asexual reproduction, standing in the middle of the road of reproductive options.

Most mosses have the ability to clone themselves from broken-off leaves or other torn fragments. These bits of debris can grow into new adults that are genetically identical to the parents, an advantage in a constant environment. The clones remain near the parents and have little ability to venture into new territory. Cloning by dismemberment may be effective but it is a decidedly crude and random way to send genes into the future. *Tetraphis* however, is the aristocrat of asexual reproduction, possessed of a beautifully sculpted design for cloning itself. When I kneel to look closely at the patches of *Tetraphis* on the old stumps, I see that the surface of the colonies is sprinkled with what look like tiny green cups. These gemmae cups, formed at the ends of the upright shoots,

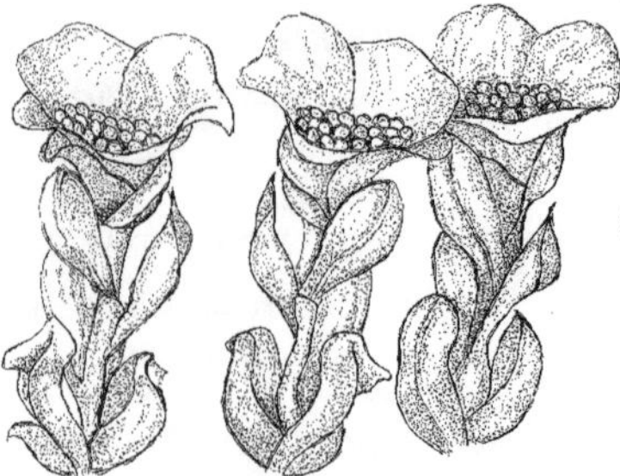

Gemmae cups of Tetraphis pellucida

resemble miniature bird's nests, complete with a clutch of tiny emerald eggs. The nest or gemmae cup is a circular bowl made of overlapping leaves and nestled within it lie the egg-like gemmae. Each gemma is a roundish mass of only ten to twelve cells, which catch the light and shimmer. Already moist and photosynthesizing, each gemma is poised to establish itself as a new plant, cloned from its parent. It rests in the nest, waiting. Waiting for an event that will propel it away from its parent, where there's room to grow and start its own family.

When the skies darken and the thunder rolls, the time is at hand. Great big raindrops pelt the forest floor, and ants and gnats dive into mosses for shelter, lest they be squashed by the momentum of the raindrops. But sturdy little *Tetraphis* waits expectantly, for it is designed to harness the power of a raindrop. When a gemmae cup receives a direct hit, the raindrop breaks loose the gemmae and propels them outward, leaving the nest empty. The gemmae can be splashed up to fifteen centimeters away, which isn't bad for a plant only one centimeter tall. In a favorable location, the gemmae can regenerate an entire new plant in the span of a single summer. In comparison to spores, which are at the mercy of a fickle breeze that deposits them anywhere, a rock or a rooftop or the middle of a lake, gemmae are more likely to land in the same neighborhood as their parents. As clonal propagules, the gemmae carry a combination of genes that has already proven successful on this stump.

In contrast, the spores produced by the sexual mixing of the parents' genes are a myriad of genetic combinations, a powder of potential sent off to seek its fortune in the unknown realm beyond the stump. There are other patches of *Tetraphis* on the very same stump which are the cinnamon color of old redwood. They take their rusty tint from

Sexual shoots of Tetraphis pellucida bearing sporophytes

the dense swath of sporophytes which rise from the green shoots below. Each sporophyte ends in a capsule shaped like an open jar. The mouth of the jar is ringed with four rusty teeth, from which the name *Tetraphis* ("four teeth") is taken. When the capsule is ripe, millions of spores will be released onto the breeze. The product of sex, the spores will carry the shuffled genes far from their parents. While these spores have the advantage of variety and distance, their success rate is exceedingly small. The tiny spores, even when carefully sown on a suitable site like another hemlock stump, yield only one plant for every 800,000 spores sown. There is clearly a tradeoff between size and success. The gemmae are hundreds of times larger than spores, and hundreds of times more effective in generating new plants. The large size and active metabolism of the gemmae, in comparison to spores, give them a higher chance of success. In experiments, I've found that one in ten gemmae survive to establish a new plant.

I can hear that the sound of the hayrake has stopped and Paulie comes down the sun-dappled path to see what I'm up to, grateful for a respite from the summer sun. I hand her my water bottle and she drinks deeply, wiping her mouth on the back of her hand and bending down to sit on a hemlock stump. I show her the two kinds of *Tetraphis*, the asexual colonies with reliable "stay at home" gemmae and the highly sexual colonies, sending their adventurous offspring off on the breeze. She just nods her head and laughs. It's a story she knows very well. Her daughter, so very like her mother, has decided to stay on and work the land alongside her parents after college. Her oldest son, however, has flown the nest to become a teacher at the other end of the state, having no interest at all in days that start with milking before sunrise and end long after the cows come home.

When I look at logs and stumps covered with *Tetraphis*, there is a striking pattern. The two forms, gemmae and spores, occur in distinct patches, almost never intermingling. Since each reproductive strategy, clonal and sexual, is usually associated with a very different environment and with individual species, I wonder at the cause of this pattern. Why should the same species adopt a clonal lifestyle in one patch and a sexual lifestyle in another on the very same stump? Why does natural selection allow two opposite behaviors to coexist in the same plant? This question led me into a long and intimate relationship with *Tetraphis*, one of fascination and of respect where *Tetraphis* taught me a great deal about doing science.

I suspected right away that the cause of the reproductive patchiness was some aspect of the physical environment. Perhaps differences in moisture or nutrients in the decaying wood caused different forms of reproduction. So I laboriously measured environmental factors to see which one was correlated with either sexual or clonal behavior. I lugged around a pH meter, a light meter, a pyschrometer, and bagged samples of decaying log to take back to the lab for an analysis of moisture and nutrients. Months of expectant data analysis later, I discovered that there was no correlation whatsoever. There seemed to be no rhyme nor reason to *Tetraphis'* reproductive choice. But if there's anything that I've learned from the woods, it's that there is no pattern without a meaning. To find it, I needed to try and see like a moss and not like a human.

In traditional indigenous communities, learning takes a form very different from that in the American public education system. Children learn by watching, by listening, and by experience. They are expected to learn from all members of the community, human and non. To ask a direct question is often considered rude. Knowledge cannot be taken; it must instead be given. Knowledge is bestowed by a teacher only when the student is ready to receive it. Much learning takes place by patient observation, discerning pattern and its meaning by experience. It is understood that there are many versions of truth, and that each reality may be true for each teller. It's important to understand the perspective of each source of knowledge. The scientific method I was taught in school is like asking a direct question, disrespectfully demanding knowledge rather than waiting for it to be revealed. From *Tetraphis*, I began to understand how to learn differently, to let the mosses tell their story, rather than wring it from them.

Mosses don't speak our language, they don't experience the world the way we do. So in order to learn from them I chose to adopt a different pace, an experiment that would take years, not months. To me, a good experiment is like a good conversation. Each listener creates an opening for the other's story to be told. So, to learn about how *Tetraphis* makes reproductive choices, I tried to listen to its story. I had understood *Tetraphis* colonies from the human perspective, as clumps in various stages of reproduction. And I had learned little by doing so. Rather than looking at the clump as an entity, I had to recognize that the clump was simply an arbitrary unit that was convenient for me, but had little meaning for the moss. Mosses experience the world as individual stems and to understand their lives I needed to make my observations at the same scale.

So I began the laborious work of inventorying the individual shoots in hundreds and hundreds of *Tetraphis* colonies. I took pains to see every patch of *Tetraphis* I sampled as a family of individuals. Every single stem was counted, and every shoot was categorized by its gender, its stage of development, and its mode of reproduction, gemmae or spores. I wonder how many shoots I've counted in all—probably millions. A dense colony of *Tetraphis* can have three hundred shoots per square centimeter. And then each colony was marked. I found that the plastic cocktail swords which impale olives in martinis make the best markers. They won't decay and the bright pink plastic makes them easy to locate the next year. And besides, I like to imagine the conversations of hikers who encounter mossy logs decorated with swizzle sticks.

The next year, I went back and found each of the marked colonies and counted them again. In notebook after notebook, I recorded the changes in their lives. And then again the year after that. Slowly, with my knees in the duff and my nose on the stump, I was starting to think like a moss.

I think that Paulie would be the first to understand this. Making a living as a dairy farmer on a few hilly acres is a tough proposition. She has been successful because she knows her herd, not as a clump, but as individuals. There's not a numbered ear tag on the farm; she knows every cow by name. She can spot when Madge is ready to calve, just by the way she walks down the hill. The time spent to know their habits and their needs gives her a competitive edge over the industrial-scale dairy farmers.

My notebooks record the fate of each patch, a changing census of the tiny moss community. With patient watching, and no direct questions, year by year, *Tetraphis* began to tell its own story. Colonies on bare wood start out with sparse and widely scattered shoots, a community with plenty of elbow room. In these low-density patches of fifty individuals in a square-centimeter sample, virtually every shoot bears a gemmae cup at its tip. The falling gemmae grow into more thrifty young shoots and by the time I return the next year the stems have gotten crowded. In colony after colony, I notice a remarkable pattern. With crowding, the gemmae disappear. There is an abrupt switch from making gemmae to making female shoots. Crowding seems to trigger the onset of sexual reproduction. With a populous colony of females and scattered males, it's not long before sporophytes appear. The colony has transformed itself from the vibrant green of gemmiferous shoots to the rusty color of spore production. When I return the next year, the colony has become even more crowded, approaching three hundred stems per square centimeter. This high density seems to trigger a radical shift in sexual expression. Now, the only shoots produced are male, with not a female or a gemmiferous shoot in sight. We discovered that *Tetraphis* is a sequential hermaphrodite, changing its gender from female to male as the colony gets crowded. This switching of gender with population density had been observed in certain fish, but never before in mosses.

In trying to piece together *Tetraphis'* story, I wanted to be sure that I understood what was going on, that the choice of having sex or making gemmae was really determined by the density of the colony. If that were true, then if I could change the density, the mosses should change behavior. Perhaps I could ask an indirect question, and perhaps they would answer. To ask the question in the language of mosses, I took a cue from Paulie's woods.

A few years ago, when she needed cash for the new heifer barn, she decided to harvest some trees from her woodlot. She shopped around carefully for a logger committed to low-impact harvest, someone who would take good care of the woods. They cut timber in winter, scattering their openings, and made a clean job of it. In the springs that followed, the thinned woods had a carpet of snowy white Trillium and yellow trout lilies blooming under the leafy canopy. The lowered density had let in more light and rejuvenated the old stand.

Like a logger in miniature, I sat poised with fine forceps over the old, dense *Tetraphis* patches. One by one, I plucked out single shoots of *Tetraphis*, stem by stem, until the density was reduced by half. And then I let them be, returning the next year to observe if they had given me an answer to my question. The unthinned patches of *Tetraphis* remained male and had started to turn brown. But the patches where I'd opened the moss canopy by thinning were green and vibrant. The holes I'd made in the *Tetraphis* turf were being filled with thrifty young shoots, bearing gemmae cups at their tips. The mosses had answered, in their own way. Low density is a time for gemmae, high density for spores.

The transformation to being male appears to have adverse consequences. Over and over, I observed that the dense male patches were starting to die back, becoming dry and brown. These tired male colonies, spent with reproduction, were then easily invaded by other mosses on the log. Sometimes, I'd find the telltale swizzle sticks in a patch where old male colonies of *Tetraphis* had disappeared, obliterated by the advance of carpet mosses. Why would *Tetraphis* adopt a sexual lifestyle that seemed to doom it ultimately to fail, headed for local extinction?

On many occasions, I'd return to a familiar stump only to find that the carefully marked patch of *Tetraphis* had vanished. In its place was a clean, bare surface of newly exposed wood. Scrambling around on my knees, I found the patch of *Tetraphis*, still impaled by its cocktail sword, at the base of the stump, where it had tumbled in a small avalanche of decayed wood. These stumps and logs were a landscape in motion. The process of decay and the activity of animals were constantly causing the logs to fall away, piece by piece. The stumps looked like small mountains, forested by mosses, with a talus slope of decayed chunks lying like fallen boulders at their base. Blocks of old wood fell away, carrying their surface cover of *Tetraphis* and creating the bare places I'd noticed. And what became of such open spaces, these patches of new wood? Looking closely, I could see that they were sprinkled with gemmae, little green eggs that had splashed into the gaps in the old *Tetraphis* cover. In the aftermath of disturbance, the seeds were sown for the next wave of *Tetraphis*.

When I stop by the barn to buy a carton of fresh brown eggs, Paulie is just coming back from a meeting. We stand there in the sun, admiring the morning glories climbing up the side of the old silo. She heard some talk of opening a casino over in the next county and we laugh about the unwary throwing their money away on chance. "Heck," she says, "we don't have to go to the casino to gamble. Farming is like blackjack, year in and year out." Milk prices are notoriously unreliable, and feed costs can triple from one year to the next. Farm income can fluctuate like clouds passing over the sun, but college tuition only goes up. That's where the Christmas trees come in, and the sheep and the feed corn. To buffer against uncertainty, Ed and Paulie run a diversified farm. The cows are the mainstay but in years when milk prices are down, maybe the lamb market will pay the kids' tuition, or maybe the Christmas trees. They survive in an era of disappearing family farms by a resilience rooted in flexibility, where stability comes from diversity.

It's the same for *Tetraphis*, a moss that is hedging its bets in an unpredictable landscape where a landslide of decay can disrupt years of steady growth. It achieves stability in an unstable habitat by freely switching between reproductive strategies. When the colony is sparse and there is lots of open space, it pays to be clonal. The gemmae can occupy the bare wood more quickly than any spore and maintain a competitive advantage against other moss species. But when it gets crowded, the only offspring that have a chance are spores. And so sexual reproduction is begun, to produce spores of divergent genetic makeup that will be blown away from the parents in their dwindling habitat. It's a gamble that any spore will land on a suitable log and be able to start a new colony. But it's a sure thing that without disturbance the colony will become extinct if it stays in one place.

The other mosses of less imaginative reproduction are slowly creeping closer, ready to engulf little *Tetraphis*. But *Tetraphis* has chosen its habitat well, taking full advantage of the rot which reliably causes disturbances to the log. Just about the time that the spent colony of *Tetraphis* is about to succumb to competitors, the face of the log peels away in a landslide of decay, exposing fresh new wood as it eliminates a patch of competitors and *Tetraphis* as well. If *Tetraphis* had to rely on spores to colonize these open spaces, its competitors would more often win the race for space. But just a few centimeters away stands a patch of *Tetraphis* in its clonal phase. With the next rain, gemmae are splashed into the opening and rapidly produce a new patch of vibrant green shoots. Decay renews the open space, and in accord, *Tetraphis* renews itself. *Tetraphis* plays both sides of the game, producing gemmae for short-term profit and spores for long-term advantage. In this changeable habitat, natural selection favors flexibility rather than commitment to a single reproductive choice. Paradoxically, those species adapted to a specialized lifestyle come and go, but *Tetraphis* persists by keeping its options open and maintaining its freedom of choice.

Maybe it's the same with our old farm, persisting now for almost two centuries. Generations of other women before us have shooed barn cats out of the road, planted lilacs, and raised their children under these maples. The old bull has been replaced by the AI man, and the cistern by a well. But the world is still unpredictable and still we survive by the grace of chance and the strength of our choices.

A Landscape of Chance

I believe it was the silence that woke me, an unnatural quiet in the silvery half-light before dawn, the hour of wood thrushes' songs. As I rose through the clouds of sleep, their absence grew alarmingly real. An Adirondack morning usually arrives to the accompaniment of veeries' and robins' songs, but not on this day. I rolled over to look at the clock. 4:15. The light outside suddenly shifted from silver to steel and thunder grumbled in the distance. The aspens turned up their leaves to flutter stiffly in the stillness, giving their rain call in the silence left by the birds. They must be hunkered down, I thought, in anticipation of the rain. Around here they say, "Rain before seven, done by eleven." I'd probably get to go canoeing after all. I snuggled back under my covers to wait it out. That's when the pressure wave hit the cabin like an axe against a tree.

Jumping out of bed, I ran to shut the cabin door, which had been suddenly flung open by the force of the wind. The cabin windows looked out onto a lake frothing and churning like the ocean, under a sky which had turned a sickly shade of green. The paper birches on the shore were bent nearly horizontal, their thrashing gyrations caught in the strobe of lightning, white on white, as a curtain of electricity advanced across the lake. The big pine over the porch began to wail and the windows seemed to press ominously inward. I herded my small daughters to the back of the cabin. We cowered in anticipation of shattered glass and splintered pine, small and speechless before the storm.

The thunder rolled and rolled, like a long freight train roaring by, then leaving silence in its wake. The sun rose over a placid blue lake. But still there were no birds. Nor would there be for the rest of that summer.

On July 15, 1996, the Adirondacks woke to a landscape battered by the most powerful storm ever recorded east of the Mississippi. Not a tornado, but a microburst, a wall of convective thunderstorms riding a pressure wave off the Great Lakes. Trees were snapped and uprooted in swaths of blowdown that took every tree. Campers were pinned in their tents and hikers stranded in the backcountry, where trails disappeared under piles of timber thirty feet high. Helicopters were dispatched to carry them to safety. In a single hour, vast tracts of shaded woodland became a jumble of torn trees and upturned soil, exposed to the glare of the summer sun.

Such land-clearing events are rare, but forests exhibit remarkable resilience in the face of disaster. I'm told that the Chinese character for catastrophe is the same as that which represents the word opportunity. And the blowdown, while catastrophic, presented opportunity for many species. Aspens, for example, are perfectly adapted to take advantage of periodic disturbances. Quick growing and short lived, aspens produce light wind-blown seeds that sail away on cottony parachutes. In order to travel fast and far, aspen seeds come with minimal baggage. They can live for just a few days, and will die unless they germinate. An aspen seed that lands on an undisturbed forest floor hasn't a chance of success. Its tiny rootlet, the key to self-sufficiency, cannot penetrate the thick leaf litter and the dense canopy shades out the sun it needs. But, in the aftermath of the storm, the forest floor has been churned up into a tumult of logs and soil thrown up by uprooted trees. In the full sun, on clean mineral soil, the aspen seedlings will be the first to colonize the devastation.

Storms such as this one come perhaps once in a century, but the wind blows nearly every day, rocking the canopy trees and weakening their hold in the soil. The predominant cause of tree mortality in the northern deciduous forest is windthrow. Gravity always wins in the end. In frequent storms, or under winter's load of ice, individual trees come crashing down with great regularity, like pendulum strokes of the ecological clock. Even on a calm day, you can sometimes hear a tree groan and lean with a whoosh to the ground. The fall of a single tree punches a hole in the canopy and a shaft of light follows it to the forest floor. These small gaps don't provide enough light for aspens to get started, but there are other species poised to take advantage of someone else's demise. Yellow birch, for example, thrives on the small mound of earth thrown up by single treefalls and quickly establishes, growing up along the column of light to meet the maples in the canopy. The mound eventually erodes away, leaving the birch standing on stilt-like roots. Yellow birch is generally considered to be a "climax" species, a member of the dominant triad of the mature beech-birch-maple forest, and yet its very presence is due to disturbance. Without treefall, yellow birch would disappear, and the triad would be diminished. Paradoxically, disturbance is vital to the stability of the forest.

A forest's resilience after disturbance lies in its diverse composition. A whole suite of species is adapted to disturbance gaps of different types. Black cherry comes in to intermediate size gaps where the soil has been exposed, hickory into small gaps on rocky soils, pine after fires, striped maple after disease. The landscape is like a partially completed jigsaw puzzle in varying shades of green, where holes in the landscape can be filled with one particular piece and no other. This pattern of forest organization known as gap dynamics is known in forests around the world, from the Amazon to the Adirondacks.

There is something reassuring about these patterns which speak of order and harmony in the way things work. But what if the forest is composed of "trees" only a centimeter tall? Are the same dynamics of gap creation and colonization also played out at a micro-scale? Do the rules for assembling the jigsaw puzzle of a landscape also apply to mosses? Part of the fascination of working with mosses is the chance to see if and when the ecological rules of the large transcend the boundaries of scale and still illuminate the behavior of the smallest beings. It is a search for order, a desire for a glimpse of the threads that hold the world together.

The trees that come crashing to the forest floor will soon become mossy logs. Like the forest above it, the moss turf is a patchwork of many species. And if you're down on your knees with your nose poked into the earthy smell, you'll see that the moss carpet is not unbroken green. There are gaps here, too, little openings in the turf where the wood shows through like bare soil after a windstorm. The dominance of the climax species is temporarily interrupted here, providing a microhabitat for any quick-thinking opportunists.

G. Evelyn Hutchinson, a pioneering ecologist, spoke eloquently of the living world as "the ecological theater and the evolutionary play." This decaying log is a stage, and the scenes take place in the gaps, where the colonists act out their drama.

Tetraphis pellucida is here, its life intertwined with the forces of disturbance. Like aspen, it cannot renew itself without space free from competition. When disturbance opens up a new gap, its gemmae are there to quickly colonize the space. As the gap gets crowded, *Tetraphis* shifts to sexual reproduction, manufacturing the spores that will carry it away to a new gap on some distant log. Spore production happens just in the nick of time, before the carpet mosses encroach on the gap and *Tetraphis* perishes beneath them. Colonizing these short-lived gaps is essential. In the absence of disturbance, *Tetraphis* would not survive.

But *Tetraphis* is not alone. The other player in the evolutionary drama is *Dicranum flagellare*. *D. flagellare* shares many traits with *Tetraphis*. It, too, inhabits decaying logs. Like *Tetraphis*, it is small, short lived, and easily outcompeted by the large carpet mosses. Like *Tetraphis*, it relies on growing in the open spaces provided by disturbance. Like *Tetraphis*, it has a mixed reproductive strategy. So here we have two species, unrelated but quite similar in their approaches to life. They occupy the same logs, at the same time, in the same forest. Ecological theory predicts that if two species are very similar competition for shared needs will eventually lead to the exclusion of one player. There will be a winner and a loser, not co-champions. How then do the two species share space on the log? How do they coexist if they are so similar? Again, theory predicts that coexistence is possible only when the two species diverge from one another in some essential way. I was intrigued by how these two gap colonists divided up the habitat. Perhaps they used parts of the gaps on the log that differed in light or temperature or chemistry. Since colonizing gaps is critical to their success, I wondered just how each of them managed to find gaps and begin new lives.

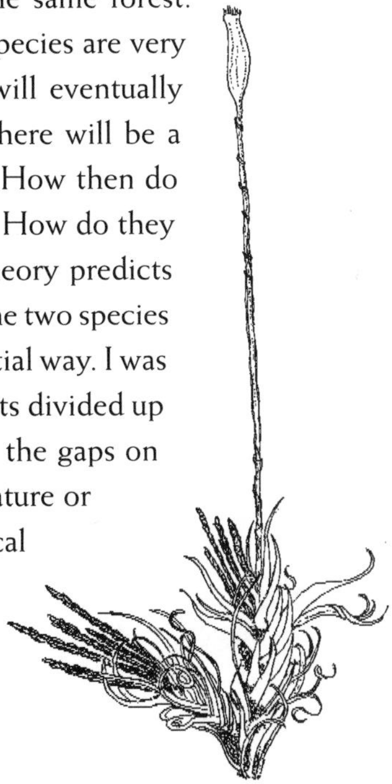

Dicranum flagellare with both sporophyte and asexual brood branches

The foliage of *D. flagellare* would never be mistaken for the round shining leaves of *Tetraphis*. Each leaf is long and stiffly pointed, like a tiny pine needle. Its reproductive strategy calls for making sexual spores as well as asexual or clonal propagules. Rather than the lovely gemmae that *Tetraphis* splashes around the log, *D. flagellare* clones itself with little bristle–like tufts at the tip of each shoot. Theoretically, these tufts can break off, releasing individual "brood branches," long thin cylinders of green, about a millimeter long. Each brood branch has the potential to clone a new plant. But potential does not always align with reality. In order to be useful, the brood branch would have to detach from the parent and somehow move to a new gap of bare wood.

Try as I might, I couldn't see how that happened. I thought they might splash like *Tetraphis* gemmae, so I set up experiments to shower them with water. Nothing. Wind? I put sticky traps around the plants to detect any brood branches that might blow from the parent. Nothing. I added a strong fan to help matters along. Still nothing. *D. flagellare* makes clonal propagules but seems incapable of using them. Non-functional parts to organisms are not uncommon. Many organisms have vestigial structures that have lost their function, like the human appendix. Perhaps brood branches were similarly useless.

My student, Craig Young, and I spent two summers on hands and knees. Dead logs and their moss communities became our world. Each gap in the mossy cover of the logs was painstakingly described. Its moisture, light, pH, size, position, the trees overhead, and the mosses at the gap edge—all was recorded in our notebooks. Contrary to popular belief, blood sacrifice did not disappear with the dawn of science. The blackflies in May, the mosquitoes in June, and the deerflies in July all benefited by the hours we spent sitting still by logs, mapping out the pieces of the jigsaw puzzle. Craig became adept at grabbing our tormentors out of the air when they were flying away, heavily laden. His notebook was splotched with squashed flies and small spatters of our blood.

Our observations revealed a pattern so clear that I marveled at its constancy. While *Tetraphis* and *D. flagellare* both colonize gaps on dead logs, there is a marked separation between them, a segregation so complete you'd think there were "*Tetraphis* only" signs posted at the gap edges. *Tetraphis* was most common in big gaps, those more than about four square inches. The bigger the gap, the more *Tetraphis*. *D. flagellare* was restricted to small gaps, generally about the size of a quarter. Because gaps occur in many shapes and sizes on the log, apparently the two species could coexist by specializing. With *Tetraphis* in big gaps and *D. flagellare* in small ones, they could avoid competition.

This pattern directly mirrored the gap dynamics of the forest overhead. *Tetraphis* responded to big gaps, as if it had taken lessons from aspen, sending out lots of highly dispersable propagules, quickly cloning itself to fill the space. *D. flagellare* seemed to be the equivalent of yellow birch, surviving by jumping into the smallest of gaps. The carpet mosses played the same role as the climax beech and maples, slow, enduring competitors poised to move in.

But the story of *Tetraphis* and *D. flagellare* is even more complex than the pattern of the trees. We found that *Tetraphis*'s large gaps and *D. flagellare*'s small ones occurred in very different places. Large *Tetraphis* gaps were nearly always on the sides of logs. *D. flagellare* was restricted to the tops of logs with striking regularity. We reasoned that the two gap sizes must have different causes, but what?

Catastrophic windstorms create opportunities for aspen, but it is fungi and the inevitability of gravity that make good *Tetraphis* habitat. In particular, a group of wood-destroying fungi known as the cubical brown rots are responsible for gap formation. These fungi digest wood in a very distinctive way, dissolving the glue between cell walls in a way that causes the wood to decay in blocks, rather than fiber by fiber as do the white rot fungi. On the steep side of a log, the decay-loosened wood needs only gravity, or the dragged hoof of a passing deer, to dislodge the blocks and send them tumbling down. The falling block may rip away a carpet of competitors, or other *Tetraphis* colonies, creating a large gap by a landslide of decayed wood.

But what of *D. flagellare*'s small gaps? Their origin remained a mystery, as did the mechanism of how their reluctant brood branches could ever detach and find their way to a waiting gap. We were missing a crucial piece of the puzzle and so, on hands and knees, we went looking.

Moist logs are prime real estate for slugs. Every morning their slime trails glisten on the mosses, their circuitous tracks like a message written on the log in disappearing ink, a script we tried to decipher with an experiment. We wondered if perhaps slugs were responsible for the movement of *D. flagellare* brood branches. We even imagined the propagules being glued to the wood with slug slime. So on misty mornings Craig and I went slug hunting. Every time we found a slug, we would gently pick it up and touch its belly to a clean glass microscope slide, like a round inky thumb pressed to a fingerprint card. Then we put the surprised slugs right back where we found them and after a moment of playing dead they continued their slow progress across the moss. With the care of detectives fingerprinting a subject, we carefully brought our slug prints back to the lab and under the microscope we checked the slime for the presence of moss propagules. Sure enough, caught in the sticky film were fragments of green. Perhaps we were on to something.

Slugs seemed to have the ability to pick up bits of moss, but could they carry it far enough to move the brood branches into the gaps? In order to gauge their potential as moss dispersers, we created a little raceway for them, a steeplechase course for mollusks. The course was a long glass plate, a smooth surface over which they could ooze with ease. We placed the freshly caught slugs at one end of the glass on a bed of *D. flagellare*, bristling with brood branches. The idea was to trace their path across the glass and measure the distances slugs carried brood branches. Craig comes from Kentucky, land of the thoroughbreds and Churchill Downs. Racing gets in your blood, I guess, and we placed bets on our favorites in the upcoming slug race and were humming "Camptown Races" as we set up the experiment. Doo-dah, doo-dah. The only problem was that the slugs were content to stay right there on the moss. They roamed a little, touched antennae, and then backed away and lay there like tiny brown walruses basking on the beach, oblivious to our expectations. Clearly, we needed something to excite them, to entice them out to glide over the glass. What is it that motivates slugs? I am an inveterate reader of garden catalogs and remembered reading that you could lure slugs out of lettuce beds at night by leaving shallow pans of beer as traps. So, using an inducement as old as civilization, we offered a refreshing beverage at the end of the race course. It worked. Antennae stretched toward the malty aroma, our subjects abandoned their sluggish behavior and headed across the glass to their reward, leaving their trails behind them.

The race was slow enough that we could go have lunch between the starting gun and the finish. As it turns out, the slugs did carry *D. flagellare* brood branches in their slime. But nearly all were dropped within only a few centimeters of the moss beds. None rode their slugs as far as the beer. Disappointed, we returned the slugs to the woods, concluding that their role in moving mosses was probably a minor one. The transport of brood branches continued to elude us.

A few days later, on a day so hot and humid we wished we had packed along some of that slug lure, we sat swatting flies by a log and eating lunch. Craig's peanut butter and jelly sandwich sat atop the log, a drip of strawberry jam running down the side. Chipmunks are bold around a field station, and are habituated to peanut butter. They practically knock on the door of the live traps, asking to be let in for a peanut butter snack, in compensation for the annoyance of being measured by a student. Tail held high, with ears alert, one came running down the log, straight for the sandwich. We looked at each other and grinned as the light bulb went on.

The next day, our *D. flagellare* steeplechase was set up again, this time in a long track, with a bed of *D. flagellare* and a volunteer chipmunk at one end and several meters of sticky white paper stretched before him. When we opened the cage door, it shot out like a bullet, ran over the moss and down the race course into another cage at the end. When we lifted it out for a close look, squirming and twisting, we found bits of green clinging to the fur of its belly and to its moist pink feet. And all along the sticky paper, for meter after meter were footprints of scattered brood branches. Eureka! Here was our distributor of brood branches, not water or wind or slugs, but a chipmunk. Its footsteps broke off the bristly brood branches, and their tiny leaves caught like burdocks on the silky fur, to be scattered along behind it. We thanked our chipmunk with great appreciation and released it back to the woods with a peanut.

Perhaps you've noticed that chipmunks, in their busy-ness rarely run on the ground. Instead, you see them following the twisted paths formed by rocks, by stumps and wood, like the "don't touch the ground" games we used to play as kids. They use logs as highways through the forest. For days, we sat quietly watching chipmunks travel along our *D. flagellare*-covered logs. Each was traversed many times a day as the chipmunks moved between feeding places and the safety of the burrow. They run with a start and a stop, sporadically putting on the brakes to do a bright-eyed check for predators. We noticed that when they came to a halt little bits of moss were kicked up from the surface, like gravel spun out by a hard-braking car. It seems that the chipmunks in their workaday lives were making the small gaps we found in the moss turf, like potholes in a roadway. And with each traverse they regularly delivered a sprinkle of *D. flagellare* propagules from their toes. Here was the missing puzzle piece. So this was why *D. flagellare* was found only on top of logs. Only where chipmunks run back and forth making opportunities for a small moss to live. How amazing to live in such a world where order arises from the seeming coincidence of the smallest things.

In time, the wind-thrown trees become mossy logs and the aftermath of the storm is a tapestry of mosses on a log, mirroring the same dynamics that shaped the forest around it. Aspen seeds flying in the wind of a tree-throwing gale create a new forest. *Tetraphis* spores spread green over a landslide gap on the side of a log. Yellow birch quietly takes its place in a single tree gap, while *D. flagellare* fills small patches on a log top. There is a home for everything, the puzzle pieces slip into place, each part essential to the whole. The same cycle of disturbance and regeneration, the same story of resilience, is played out at a minute scale, a tale of the interwoven fates of mosses, fungi, and the footfalls of chipmunks.

City Mosses

I f you're a city dweller, you don't have to go on vacation to see mosses. Sure, they're much more abundant on a mountaintop or in the falls of your favorite trout stream, but they also live alongside us every day. The city mosses have much in common with their urban human counterparts; they are diverse, adaptable, stress-tolerant, resistant to pollution, and thrive on crowded conditions. They are also well traveled.

A city offers mosses a multitude of habitats which may otherwise be quite uncommon in nature. Some moss species are far more abundant in the human-made environment than they are in the wild. *Grimmia* doesn't discriminate between a granite crag in the White Mountains and a granite obelisk on Boston Common. Limestone cliffs are not abundant in nature, but there's one on every Chicago street corner and mosses perch contentedly on its pillars and cornices. Statues provide all kinds of water-holding niches where mosses abound. Next time you walk through the park, look in the folds of the flowing coat of whatever general sits mounted on a pedestal, or in the wavy marble locks of Justice's hair outside the courthouse. Mosses bathe at the edges of our fountains and trace the letters on our gravestones.

Ecologists Doug Larson, Jeremy Lundholm, and colleagues have speculated that the stress-tolerant, weedy species that cohabit our urban spaces may have been with us since our earliest days as a species. In their urban cliff hypothesis, they note a striking number of parallels between the flora and fauna of natural cliff ecosystems and the vertical walls of cities. Many weeds, mice, pigeons, house sparrows, cockroaches, and others are all endemic to cliff and talus-slope ecosystems, so perhaps it is

Cushion of Grimmia pulvinata

no surprise that they willingly share our cities. The same can be said for urban mosses, many of which are typical of rock outcrops, whether natural or human-made. We tend to devalue the flora of cities as a depauperate collection of stragglers, arising *de novo* with the relatively recent development of cities. In fact, the urban cliff hypothesis suggests that the association between humans and these species may be ancient, dating from our pre-Neanderthal days when we both took refuge in cave and cliff dwellings. In creating cities, we have incorporated design elements of the cliff habitat and our companions have followed.

Admittedly, city mosses are not the soft feathery mats of forest mosses. The harsh conditions of urban life limit them to small cushions and dense turfs as tough as the places they inhabit. The arid conditions of pavements and window ledges cause mosses to dry out quickly. To defend against drying, the moss shoots pack closely together, so that limited moisture may be shared among shoots and held as long as possible. *Ceratodon purpureus* makes such dense colonies that when dry they resemble small bricks; when wet, green velvet. You'll find *Ceratodon* most commonly in gravelly spots, like at the edge of a parking lot or on a rooftop. I've even seen it growing on the rusted metal of old Chevys and abandoned railroad cars. Every year it produces a dense crop of unmistakable purplish sporophytes to send its spores off to the next bare patch.

The most ubiquitous of mosses, urban or otherwise, is *Bryum argenteum*, the Silvery Bryum. I have never traveled without encountering *Bryum* on my journey. It was on the tarmac in New York City and on the tiled roof outside my window in Quito the next morning. *Bryum* spores are a constant component of aerial plankton, the cloud of spores and pollen which circulates all around the globe.

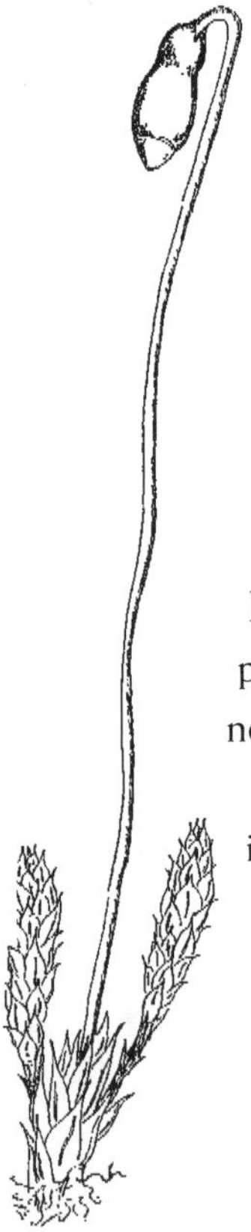

Shoot and sporophyte of Bryum argenteum

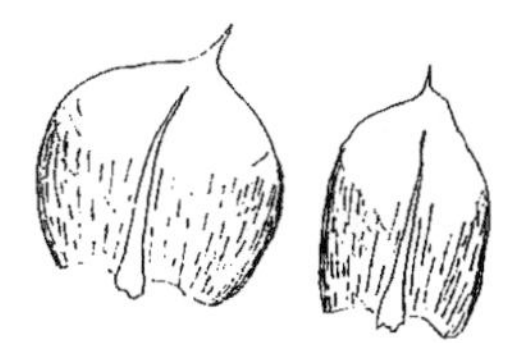

Leaves of Bryum argenteum

You've probably walked over millions of *Bryum* without ever realizing it, for it is the quintessential moss of sidewalk cracks. After a rain, or a hosing down by a sanitation worker, water lingers in the tiny canyon of a fissure in the pavement. Mingling with the nutrients provided by the flotsam of pedestrians, the crack becomes ideal for Silvery Bryum. It takes its name from the burnished silver color of the dry plants. Each tiny round leaf, less than a millimeter long, is fringed with silky white hairs, visible with a magnifying glass. The shiny hairs reflect away the sun and protect the plant from drying. Under the right conditions, the pearly plants will put up a host of sporophytes to cast their young into the aerial plankton, so that a New York *Bryum* could easily end up in Hong Kong. However, the much more common route for dispersal is by footsteps. *Bryum* shoots are fragile at their tips and in fact are designed to break off. The broken tips, scuffed along by a pedestrian, will take hold in another sidewalk, spreading Silvery Bryum all over the city.

The native habitat of *B. argenteum* is quite specialized and finds many equivalents in an urban setting. It has doubtless become far more abundant with the rise of cities than it was in our agrarian past. For example, the natural habitat of *Bryum* includes seabird rookeries where it colonizes the accumulated guano. Its urban counterpart is a pigeon-stained windowsill where it will form silvery cushions among the droppings. Similarly, *B. argenteum* is associated with prairie dogs in the Midwest and lemmings in the Arctic, where it spreads out like a welcome mat at the entrance to their burrows. The animals urinate by their doorways to mark their territory, and *B. argenteum* thrives where nitrogen is abundant. The base of a city fire hydrant is equally inviting.

Lawns are another good place to look for mosses—if you have a chemical-free lawn, that is. Down at the base of the grass plants are often threads of *Brachythecium, Eurhynchium,* or any number of other species, trailing among the grasses.

One of the pleasures of university life is fielding questions from the community about biology. Sometimes people send in plants to be identified or ask about the use of a certain plant. But I'm saddened that many requests we get concern how to kill something. A soil ecology colleague tells the story of a panicked phone call from a woman who had followed the directions in a pamphlet he'd written to start a compost pile in her backyard. Several weeks had elapsed and when she checked the pile of leaves and salad scraps, she was horrified to find that it was full of bugs and worms. She wanted to know how to kill them.

I once had a phone call from an urban homeowner asking for information on how to kill the mosses in his lawn. He was quite convinced that the mosses were killing his carefully tended lawn and wanted revenge. I asked a few questions and learned that the grass was on the north side of the house in the deep shade of maples. What my caller was observing was that his grass was dwindling, and the mosses which had always been there were now taking advantage of the open space. Mosses cannot kill grasses. They simply haven't the ability to outcompete them. Mosses appear in a lawn when conditions for moss growth are better than conditions for grass growth. Too much shade or water, too low a pH, soil compaction—any of these things can discourage grasses and let the mosses appear. Killing the mosses would not help the ailing grass in any way. Better to increase the sunlight, or better yet, pull out the remaining grass and let nature build you a first-rate moss garden.

The abundance of mosses in a city owes much to the local rainfall. Seattle and Portland support the most prolific urban moss floras I know. It's not just the trees and buildings—in the long rainy winters moss will grow on just about anything. I used to walk past a fraternity house at Oregon State University where a tree was festooned with shoes, high in the branches. From time to time, a shoelace would rot through and a sneaker would plummet to the sidewalk, completely engulfed in moss.

Oregonians seem to have a love-hate relationship with their mosses. On the one hand, there's a certain civic pride among folks who label themselves mossbacks and cheer for teams with aquatic mascots—Beavers and Ducks. On the other hand, moss eradication is big business. Hardware stores stock shelves of chemicals with names like Moss-Out, Moss-B-Gone, and X-Moss. An ad campaign on Portland billboards read: "Small, Green, Fuzzy? Kill It!" These chemicals eventually end up in the streams and in the food chain of threatened salmon. And the mosses always come back. Roofing professionals have led homeowners to believe that mosses lead to degradation of the shingles and eventually to leaks. For an annual fee, they will gladly remove them. Allegedly, the moss rhizoids penetrate tiny cracks in the shingles and accelerate their deterioration. However, there is no scientific evidence to support or refute this claim. It seems unlikely that microscopic rhizoids could pose a serious threat to a well-built roof. One technical representative for a shingle company acknowledges that he's never seen any damage by mosses. Why not just let them be?

In such a utopian climate for mosses, living roofs seem an ideal alternative to endless extermination. A mossy roof can actually protect shingles from the cracking and curling caused by intense exposure to the sun. Moss adds a cooling layer in the summer and, when the rains come, slows stormwater runoff. And besides, a mossy roof is a thing of beauty. Golden cushions of *Dicranoweisia* and thick mats of *Racomitrium* are much more appealing than a barren expanse of asphalt shingles. And yet we invest lots of time and money in removing them. There seems to be an unspoken agreement in a trim suburban neighborhood that a mossy roof represents a hint of moral decay as well as decomposing shingles. The ethics seem inverted. A mossy roof has come to mean that the homeowner is somehow negligent in his/her responsibilities for maintenance. Shouldn't the moral high ground belong to the folks who've found a way of living with natural processes rather than battling them? I think we need a new aesthetic that honors a mossy roof as a status symbol of how responsibly the homeowner behaves in maintaining the ecosystem. The greener the better. Neighbors would look askance at the owner of a roof scraped bare of friendly moss.

Some city folks try to get rid of mosses, and some invite them in. The most remarkable assemblage of urban mosses I've ever seen inhabits a loft in Manhattan. I usually hike or paddle a canoe to see my favorite mosses, but this time I took a subway and eventually an elevator to the fifth floor high above the streets of New York City, to the home of Jackie Brookner. Jackie is small and quiet, but she has a gleam that makes her stand out in the crowd, like a richly colored pebble on a gravel beach. I went to see her because that summer we were both working on boulders.

My boulder is Adirondack anorthocite, rolled twelve thousand years ago to the shores of Whoosh Pond by a glacier. Her boulder began as an aluminum armature with a skin of fiberglass cloth stretched over its contours. She mixed sand and gravel into cement and with her hands formed ridges and valleys over the surface. She then pressed soil into the still-wet surface. My boulder is lit by sunflecks through a canopy of maples and moistened by an overnight rain and the mists from the stream where brook trout lie in the shadows. Her boulder is illuminated by a bank of growlights suspended from the high ceiling of the loft and watered by a spray system on a timer. It sits in a blue plastic wading pool where goldfish hide beneath the lily pads. My boulder is named #11N. Her boulder is named Prima. It's short for Prima Lingua, the First Tongue.

Jackie is an environmental artist. Her loft is full of ideas made visible: chairs shaped from earth, nests of roots and wire, and a collection of feet, sharecroppers' feet, molded from the same native clay that lay beneath their cotton. Prima Lingua: the first tongue speaks in the first language, the sound of water flowing over rock. Prima's looming presence—she stands six feet tall—also speaks of environmental processes, the cycling of water and nutrients, and the interconnectedness of the animate and inanimate worlds. Jackie's creation is more than "rock" and water, it is a living boulder covered with mosses. The prepared surface was first inoculated by the moss spores which flew in her open window above the streets of Manhattan. *Bryum argenteum* and *Ceratodon purpureus* were among her first colonists. Mosses and rocks were meant to be together, no matter their origins. On walks and travels, Jackie also picked up bits of moss and invited them home to live with Prima. A thriving community began to form when the right conditions were created.

Prima is also about ecological restoration. Its beauty is as much functional as visual. The living sculpture is actively purifying the water which flows over it. Mosses have an exceptional capacity for removing toxins from water, binding them to cell walls. Jackie's artwork is being explored for use in wastewater treatment and protection of urban streams.

Together, we pore over Prima with magnifying glasses, looking at the species patterns and the mites and springtails which move among the leaves. Jackie's art materials are protonema and sporophytes and she knows them well. A small microscope sits on the same table with sketches and inks. Drawings of archegonia are taped above her work table. The sad truth is that many scientists believe they have the sole method for understanding the workings of the natural world. Artists don't seem to share that illusion of exclusive truth. In midwifing the birth of a moss community, Jackie has discovered more about moss establishment on rock than any scientist I know. We stay up half the night talking, with Prima murmuring agreement in the background.

Amidst the traffic and the smokestacks, city dwellers confront the health impacts of air pollution every day. When you draw a breath of air, it is pulled deep and then deeper into your lungs. Down tiny branching pathways, closer and closer to the bloodstream which is waiting for the oxygen it carries. In the alveoli, your breath is but a single cell away from your blood. The cells are glistening and wet, so that the oxygen may dissolve and pass over. Through this thin watery film, deep in the lungs, our bodies become continuous with the atmosphere. For better and for worse. The urban epidemic of asthma is symptomatic of a wider air-quality problem. The health of mosses in a neighborhood also reflects the level of air quality. Mosses and lichens are both very sensitive to air pollution. Street trees which once were greened over by moss are now bare. Check out the trees in your neighborhood. Their presence or absence has meaning. They are the canaries in the mine.

Mosses are much more susceptible to air pollution damage than are higher plants. Of particular concern is the sulfur dioxide which spews from power plants. It is a by-product of combustion of high-sulfur fossil fuels. The leaves of grasses, shrubs, and trees are many layers thick and are coated with a waxy layer, the cuticle. Mosses have no such protection. Their leaves are only a single cell thick, so, like the delicate tissue of your lung, they are in direct contact with the atmosphere. This is advantageous in clean air, but disastrous in areas polluted with sulfur dioxide. A moss leaf has much in common with the alveolus, it works only when it is wet. The water film allows the beneficial gases of photosynthesis, oxygen and carbon dioxide, to be exchanged. However, when sulfur dioxide meets that water film it turns to sulfuric acid. Nitrous oxide from car exhausts turns to nitric acid, and also bathes the leaf in acid. Without the protection of a cuticle, leaf tissue dies and becomes bleached and pale. Eventually, most mosses are killed by such severe conditions, leaving polluted urban centers virtually without mosses. Mosses began disappearing from cities soon after industrialization began and continue to decline wherever air pollution is serious. As many as thirty species which once flourished in cities have all but vanished, as air pollution increased.

The sensitivity of mosses to air pollution makes them useful as biological monitors of contamination. Different moss species are tolerant of varying levels of pollution in highly predictable ways. The type of mosses present on a tree can be used as a measurement of air quality. For example, the presence of *Ulota crispa* in dime-sized domes on a tree indicates that sulfur dioxide levels are less than 0.004 ppm, since it is highly sensitive to pollution. Urban bryologists have observed that the moss flora changes in concentric zones, radiating outward from the city center. Mosses are often absent at the center, but several tolerant species inhabit the next zone, with increasing numbers of species at the margins of the city. The good news is that when air quality improves, the mosses return.

Some people, myself included, could never live in a city. I go to the city whenever I must and leave as soon as I can. Rural folks are more like *Thuidium delicatulum*. We need a lot of room and shady moisture to flourish, choosing to live along quiet brooks rather than busy streets. Our pace of life is slow and we are much less tolerant of stress. In a city, that lifestyle would be a liability. On the streets of New York City the *Ceratodon* style is much in demand, fast paced, always changing, and making the best of the crowds. The urban landscape is not the native habitat for mosses or for humans and yet both, adaptable and stress tolerant, have made a home

Shoot of Ulota crispa

there among the urban cliffs. Next time the bus is late, take those waiting minutes to look around for signs of life. Mosses on the trees are a good sign, their absence a concern. And everywhere beneath your feet is *Bryum argenteum*. Amidst the noise and the fumes and the elbowing crowds, there is some small reassurance in the moss between the cracks.

The Web of Reciprocity: Indigenous Uses of Moss

With the first scent of burning sage, the ripples on the surface of my mind become still and it is as if I am looking deep into clear sunlit water. Murmured prayer surrounds me with wisps of smoke and I can hear each word inside me. My uncle Big Bear smudges us in the old way, calling upon the sage to carry his thoughts to the Creator. The smoke of our sacred plants is thought made visible and his thoughts are a blessing breathed in.

Big Bear's voice is low; he's tired from a day driving into the city where he's been negotiating to obtain an old school building, abandoned in the remote foothills of the Sierra. I admire the way he walks in both worlds, that of government red tape and the traditional ways. His vision is to start a new kind of school for kids in the area. His school would teach the fundamentals. How to read a river in order to catch a fish, how to gather food plants, how to live in a way that is respectful of those gifts. He values a modern education and is proud of his grandsons' straight A's. But, in his work with troubled families he sees every day the costs of not learning about respectful relationship.

In indigenous ways of knowing, it is understood that each living being has a particular role to play. Every being is endowed with certain gifts, its own intelligence, its own spirit, its own story. Our stories tell us that the Creator gave these to us, as original instructions. The foundation of education is to discover that gift within us and learn to use it well.

These gifts are also responsibilities, a way of caring for each other. Wood Thrush received the gift of song; it's his responsibility to say the evening prayer. Maple received the gift of sweet sap and the coupled responsibility to share that gift in feeding the people at a hungry time of year. This is the web of reciprocity that the elders speak of, that which connects us all. I find no discord between this story of creation and my scientific training. This reciprocity is what I see all the time, in studies of ecological communities. Sage has its duties, to draw up water to its leaves for the rabbits, to shelter the baby quail. Part of its responsibility is also to the people. Sage helps us clear our minds of ill thoughts, and carry our good thoughts upward. The roles of mosses are to clothe the rocks, purify the water, and soften the nests of birds. That much is clear. I'm wondering though, what is the gift they share with the people?

If each plant has a particular role and is interconnected with the lives of humans, how do we come to know what that role is? How do we use the plant in accordance with its gifts? The legacy of traditional ecological knowledge, the intellectual twin to science, has been handed down in the oral tradition for countless generations. It passes from grandmother to granddaughter gathering together in the meadow, from uncle to nephew fishing on the riverbank, and next year to the students in Big Bear's school. But, where did it first come from? How did they know which plant to use in childbirth, which plant to conceal the scent of a hunter? Like scientific information, traditional knowledge arises from careful systematic observation of nature, from the results of innumerable lived experiments. Traditional knowledge is rooted in intimacy with a local landscape where the land itself is the teacher. Plant knowledge comes from watching what the animals eat, how Bear harvests lilies and how Squirrel taps maple trees. Plant knowledge also comes from the plants themselves. To the attentive observer, plants reveal their gifts.

The sanitized suburban life has succeeded in separating us from the plants that sustain us. Their roles are camouflaged under layers of marketing and technology. You can't hear the rustle of corn leaves in a box of Froot Loops. Most people have lost the ability to read the role of a medicine plant from the landscape and read instead the "directions for use" on a tamper-proof bottle of Echinacea. Who would recognize those purple blossoms in this disguise? We don't even know their names anymore. The average person knows the name of less than a dozen plants, and this includes such categories as "Christmas Tree." Losing their names is a step in losing respect. Knowing their names is the first step in regaining our connection.

I was so lucky. I grew up knowing plants, wandering the fields and staining my fingers red with tiny wild strawberries. My baskets were pretty crude, but I loved gathering the willow shoots and soaking them in the creek. My mother taught me the names of plants, and my father which trees made the best firewood. When I went away to college to study botany, the focus shifted. I learned all about plant physiology and anatomy, habitat distributions, and cell biology. We carefully studied plant interactions with insects, with fungi, and with wildlife. But I don't think a word was ever said about people. Especially not Native people, even though our college sits on the ancestral homelands of the Onondaga, the center of the great Iroquois Confederacy. Humans were carefully excluded from the story, either by accident or by design, I'm not sure. I got the impression that the stature of science would somehow be lessened if we included human relationships. So when Jeannie asked me to be a partner in guiding plant walks at Onondaga Nation, I was at first reluctant. I admitted with regret that all I could offer were names and explanations of ecology. I found that Jeannie valued the scientific way of knowing that I could bring to our class, but of course, I ended up learning much more than I taught.

I have been blessed with good teachers. I am grateful for the guidance of my friend and teacher, Jeannie Shenandoah, a traditional Onondaga herbalist and midwife. There is solidity about her, she moves as if she is aware of the ground beneath her feet. We grew to share a wonderful partnership in our teaching. I'd contribute whatever I knew about the biology of the plants we found, and she would share her knowledge of traditional uses. Walking beside her, clipping twigs of crampbark for childbirth, poplar buds for salves, I began to understand the woods in a different way. I had studied with fascination the intricate connections between plants and the rest of the ecosystem. But the web of interconnection had never before included me, except as an observer, outside looking in. Then from Jeannie I learned to treat my daughter's cough with syrup from the black cherry on my hilltop, and lower a fever with boneset collected from the edge of my pond. As I gathered greens for dinner I regained my childhood relationship with the woods, one of participation, of reciprocity, and thanksgiving. It's just about impossible to feel academic detachment from the land with a bellyful of wild leeks, fragrant, hot, and buttered.

I have been wrapped up in the lives of mosses for lots of years, but I understand that our encounters had been at arm's length. We met on an intellectual plane. They teach me about their lives, but our lives have not been joined. To really know them, I need to know what role was given them when the world was beginning. What did the Creator whisper to them about their gift in caring for people? I asked Jeannie about how her people had used mosses and she didn't know. They weren't used as a medicine or a food. I know that mosses must be a part of this web of reciprocal relationship, but generations removed from the immediate connection, how are we to know? Jeannie showed me that the plants still remember, even when the people have forgotten.

In traditional ways of knowing, one way of learning a plant's particular gift is to be sensitive to its comings and goings. Consistent with the indigenous worldview that recognizes each plant as a being with its own will, it is understood that plants come when and where they are needed. They find their way to the place where they can fulfill their roles. One spring Jeannie told me about a new plant that had appeared along the old stone wall in her hedgerow. Among the buttercups and mallows was a big clump of blue vervain. She'd never seen it there before. I offered up some explanation about how the wet spring had changed the soil conditions and made way for it. I remember how she raised a skeptical eyebrow, but respectfully did not correct me. That summer, her daughter-in-law was diagnosed with liver disease. She came to Jeannie for help. Vervain is an excellent tonic for the liver and it was waiting in the hedgerow. Over and over again, plants come when they are needed. Is there something in this pattern that can tell us anything about how mosses were used? They occur everywhere, as part of the everyday landscape, so small that they often escape our notice. In the language of plant signs, perhaps this speaks of their role in human households, a small and unobtrusive role. It's the small everyday items we miss the most when they are gone.

I asked Big Bear and other elders what they could tell me of moss use and found nothing. There are too many generations and too much government-sponsored assimilation between the elders today and those who used the mosses. So much has been lost through disuse. So like any good academic I went to the library. I pored over the archived field notes of anthropologists to forage for old connections to mosses, reading old ethnographies to try and glean a hint of what the old ones would say if I could only ask. I hoped these pages would be like the sage smoke, their thoughts made visible.

I take great pleasure in gathering plants, filling my basket with roots and leaves. Usually I go with a specific plant in mind, when it's time for elderberries or the bergamot is heavy with oils. But it's the wandering itself that has such appeal, the unexpected discoveries while looking for something else. I get the same feeling in the library. It's so very much like picking berries—the peaceful field of books, the concentrated attention of the search, and the knowledge that hidden somewhere in the thicket is something worth finding.

I sifted through dictionaries of native languages, looking to see if there were indigenous words recorded for moss. I assumed that if moss was part of the everyday vocabulary, then it was also part of everyday use. In obscure proceedings of various academic societies, I found not one word, but many. Words for moss, for tree moss, berry moss, rock moss, water moss, and alder moss. The English dictionary on my desk has only one, reducing the 22,000 species to a single type.

While mosses live in every habitat, and are named by the people, I'm finding scarcely a trace of them in the transcribed notes from anthropologists. Maybe they played so small a role that their presence was scarcely worth reporting. Or maybe the reporters didn't know enough to ask. For example, I'm finding accounts of building homes, from longhouses to wigwams, replete with construction details on the way that planks were hewn and bark shingles applied. There is hardly a mention that mosses were used to chink the cracks between the logs. That's not very noteworthy until the winter wind comes rushing in. An icy wind at the back of your neck does tend to grab your attention.

The insulating nature of packed moss was also good for keeping winter cold away from fingers and toes. In browsing through source after source, I find that northern people traditionally lined their winter boots and mittens with soft mosses for an extra layer of insulation.

When the renowned "Ice Man," a 5,200-year-old body from a melting Tyrolean glacier, was recovered, his boots were found to be packed with mosses, including *Neckera complanata*. The moss actually provided an important clue as to his origins, since *Neckera* was known to occur only in lowland valleys, some sixty miles to the south. In the boreal forest, where feather mosses are a blanket beneath the spruces, their warm cushioning was also put to good use in bedding and pillows. Linnaeus, the "father of modern plant taxonomy," reports sleeping on a bedroll of portable *Polytrichum* moss, as he traveled among the indigenous Sammi peoples of Lapland. A pillow made of *Hypnum* mosses was said to impart special dreams to the sleeper. In fact, the genus name *Hypnum* refers to this trancelike effect.

I can glean that mosses were woven as decoration into baskets, used as lamp wicks and for scrubbing dishes. I'm pleased to have discovered these small notes that show that people were not oblivious of mosses, that they did play a role in daily life. But I'm also disappointed. There is nothing here that speaks of a special gift from the Creator, a unique role that could be fulfilled by no other plant. After all, dry grass can also insulate boots and a layer of pine needles can make a soft bed, too. I was hoping to find a use that reflects the essence of mossness. I was hoping to find that the people of that distant time knew mosses the way I do.

The library brought me a little further, but intuition told me that the story found there was incomplete. Every way of knowing has its own strengths and weaknesses. Taking a breather, hidden behind the accumulated stacks of books, I remembered going with Jeannie to look for plants just as soon as the snow melted and green shoots started to poke up through the winter's matted leaves. One of the first plants we found in bloom was coltsfoot, growing along the gravelly bank of Onondaga Creek. A botanist might explain this preference for March streambanks by its physiologic requirements, or perhaps its intolerance of competition. This is quite probably true. However, in the Onondaga understanding, coltsfoot grows here because it is near to its use; the medicine lies close to the source of the illness. After a long winter, just after the ice goes out, the running water is irresistible to kids. They wade and splash and race sticks in the current, soaking themselves, oblivious to the deep chill until they get home and wake up coughing in the night. Coltsfoot tea is good for just that kind of cough, the kind that comes with wet feet in small children. Another tenet of indigenous plant knowledge is that we can learn a plant's use by where it occurs. For example, it's well known that a medicinal plant frequently occurs in the vicinity of the source of the illness. There's nothing in Jeannies' telling that negates the scientific explanation. It expands the question beyond how coltsfoot lives beside the creek, to the question of why, crossing over a boundary where plant physiology cannot follow.

The plant's purpose can be read through its place. I remember this when I'm tromping through the woods and mistakenly grab a vine of poison ivy to haul myself up a steep bank. I look immediately for its companion. Remarkable in its fidelity, jewelweed is growing in the same moist soil as the poison ivy. I crush the succulent stem between my palms with a satisfying crunch and a rush of juice, and wipe the antidote all over my hands. It detoxifies the poison ivy and prevents the rash from ever developing.

So, if plants show us their uses by where they live, what is the message from mosses? I think of where they live, in bogs, along streambanks, and in the spray of the waterfalls where salmon jump. And if this weren't sign enough, they reveal their gifts every time it rains. Mosses have a natural affinity for water. Watch a moss, dry and crisp, swell with water after a thunderstorm. It's teaching its role, in language more direct and graceful than anything I've found in the library.

Perhaps the limited information on mosses in nineteenth-century anthropology is rooted in the fact that most of the observers of indigenous communities were upper-crust gentlemen. They focussed their studies on what they could see. And what they could see was conditioned by the world they came from. Their notebooks bulged with records of the pursuits of men; hunting, fishing, and making tools. When moss once appeared in a weapon, as wadding behind a harpoon tip, it was described in considerable detail. Then, just at the point when I'm ready to give up the search, I find it. A single entry. You can almost see the blush in the brevity of the statement: "Moss was in widespread use for diapers and sanitary napkins."

Imagine the complex relationships that lie behind that one entry, reduced to a single sentence. The most important uses of mosses, roles that reflect their best gifts, were everyday tools in the hands of women. Somehow I'm not surprised that the gentlemen ethnographers did not delve into the details of baby care, particularly the unglamorous but inescapable issue of diapers. And yet what could be more fundamental to the survival of a family than the well-being of babies? In this time of disposable diapers and antiseptic baby wipes, it's hard to envision infant care without this technology. If I try to imagine carrying an infant on my back all day without benefit of diapers, I don't like the image that comes to mind. I know with certainty that our grandmothers' grandmothers would have figured out an ingenious solution. In this most fundamental aspect of family life, mosses showed their great utility. To say nothing of humility. Babies were packed in their cradleboards in a comfy nest of dried moss. We know that *Sphagnum* moss can absorb twenty to forty times its weight in water. This rivals the performance of Pampers, making it the first disposable diaper. A pouch filled with mosses was probably as vital to those mothers as is the ubiquitous diaper bag today. The plentiful air spaces in dried *Sphagnum* would wick the urine away from the baby's skin, just as it wicks up moisture in a bog. The acid astringency and mildly antiseptic properties even prevented diaper rash. Like the coltsfoot, the spongy mosses placed themselves near at hand, right at the edge of the shallow pools where mothers knelt to wash their babies. They came where they were needed. As a mother at the beginning of a new millennium, I feel a certain regret that my babies never felt the touch of soft moss against their skin, forging a bond with the world that Pampers can never provide.

A woman's life was also intertwined with mosses during her menstrual period, known as her "moontime" in many traditional cultures. Dry mosses were widely used as sanitary napkins. Again, the ethnographic information is sketchy here, as males were not privy to the activities of women in menstrual seclusion huts. I imagine the huts as gathering places for the women in synchronous moontime, which occurs in communities subject to night skies uninterrupted by artificial light. The conventional wisdom of anthropologists is that menstruating women were isolated from daily life because they were unclean. But this interpretation grew from the cultural assumptions of the anthropologists and not from indigenous women themselves, who tell a different story. Yurok women describe a time of meditation and speak of special mountain pools where only moontime women were permitted to bathe. Iroquois women tell that any prohibitions on women's activities in their moontime arose because women were at the height of their spiritual powers at this time, and the powerful flow of energy could disrupt the balance of energy around them. In some tribal people, menstrual seclusion was a time of spiritual purification and training, akin to the sweat-lodge training of men. Tucked among the objects in their huts must have been baskets of mosses, selected with great care for their purpose. It seems an inescapable conclusion that women were skilled observers of different moss species, knowing their texture and creating an intimate taxonomy long before Linnaeus. The good missionary ladies must have grimaced in horror at this practice, but I think something was lost in the transition to boiled white rags.

I find another ethnography, this one written by a woman, Erna Gunther. It is full of observations of the work of women, particularly food preparation. Mosses themselves were not used for food. I've tasted them and one bitter gritty taste will dispel any thought of a meal of mosses. But while they were not eaten directly, they were an important part of food preparation among the tribes of the rainy Pacific Northwest where mosses are especially abundant. The two staple foods in the watershed of the Columbia River are salmon and camas root, both of which are revered for their gifts of sustaining the people and both of which are connected to mosses.

Salmon harvesting is generally an activity that requires the contribution of the entire family. Fishing itself is the province of the men and the women prepare the fish for drying over an alderwood fire. The dried smoked salmon will feed the tribe throughout the year and the process must be done carefully in order to assure the quality and the safety of the food. Prior to drying, the slimy coating on the newly caught fish must be wiped away. This removes potential toxins and keeps the fish from shriveling up when it is dried. In early days, salmon wiping was done with mosses. Ethnographies of the Chinook-speaking peoples describe how women would store large quantities of dry moss in boxes and baskets, to have an ample supply on hand when the salmon were running.

Mosses play a supporting role in another staple food of the Northwest, camas. Camas (*Camassia quamash*) is a member of the lily family and produces a spray of royal blue flowers in the spring. The wet meadows where it occurs were carefully tended by the tribes, including the Nez Perce, the Calapooya, and the Umatilla. Careful tending by burning, weeding, and digging produced large camas prairies. Lewis and Clark reported expanses of blooms so huge that from a distance they mistook the camas swales for shimmering blue lakes. The expedition had survived a difficult crossing of the Bitterroot Mountains and was near starvation. The Nez Perce fed them on their winter stores of camas and saved their lives.

The underground bulb is starchy and crisp, tasting somewhat like a raw potato. It is usually not eaten fresh, but painstakingly prepared by a method that yields a thick paste with the sweetness of molasses. The camas was prepared by creation of a pit oven for baking and steaming. The earthen pit was lined with hot rocks and layers of camas bulbs were placed in the pit. A mat of wet moss was then laid over the camas, building up a stack of alternating layers of moss and camas. The entire oven was topped with ferns and a fire built over the top, which burned all night long. The wet mosses provided a source of steam, which permeated the camas bulbs and baked them to a deep brown color. When the oven was opened and cooled, the steamed camas was shaped into loaves or bricks for storage. Camas was consumed all year round and traded widely throughout the west, packaged in a wrapping of moss and ferns.

Camas remains an honored ceremonial food among the western tribes, even today. At Onondaga in upstate New York, the year is marked by ceremonies of thanksgiving to the plants, as they appear each in their turn, first the maples, then the strawberries, the beans, and the corn. October at Big Bears in California brings a feast to acknowledge the acorns. As far as I know, there is no special ceremony for mosses. Maybe it's more fitting to honor these small everyday plants in small everyday ways. Cradling our babies, holding our blood, stanching a wound, keeping out the cold—isn't this the way we find our place, by participation in the life of the world?

The people gather together to give thanks that the plants, the grand and the humble, have once again fulfilled their caregiving responsibilities to the people. Tobacco will be burned in their honor. In my culture, tobacco is a bringer of knowledge. I think it's also important that we honor the different paths that lead to knowledge, the teachers in the oral tradition, the teachers in the written tradition, and the teachers among the plants. It's the time we should also turn our thoughts to our own responsibilities. In the web of reciprocity, what is our special gift, our responsibility that we offer to the plants in return?

Our ancient teachers tell us that the role of human beings is respect and stewardship. Our responsibility is to care for the plants and all the land in a way that honors life. We are taught that using a plant shows respect for its nature, and we use it in a way that allows it to continue bringing its gifts. The role of our sacred sage is to make thoughts visible to the Creator. We can learn from this teacher and live in such a way that our thoughts of respect and gratitude are also made visible to the world.

The Red Sneaker

As I dance alone in a sunlit bog, the ground beneath my feet rolls in slow waves. For a long seasick moment my foot hangs in midair, waiting for a solid place to stand. Every step sets off a new undulation, like walking on a waterbed. I reach out to steady myself, grabbing a branch of tamarack, but I've stood too long in one place and the cold water rises around my ankle. The bog sucks at my foot and I drag it out with a slow sucking sound, my leg coated to midcalf in black muck. I'm glad I left my boots back at the top of the esker. An old red sneaker of mine lies somewhere in the depths, lost on another research trip several years ago. Now I go barefoot. Apart from its proclivity for stealing footwear, a quaking bog is a lovely place to spend an August afternoon.

The ring of trees at the perimeter walls off the bog from the rest of the forest. The circle of *Sphagnum* glows green like a firefly against the wall of dark spruce. Here, the seen and the unseen worlds our elders speak of coexist in close proximity, the sunlit surface of the bog and the dark depths of the pond. There is more here than meets the eye.

The lands of my ancestors are dotted with kettle hole bogs among the forests of the Great Lakes. The Anishinabe people conduct their ceremonies using the Water Drum, a drum so sacred that it is not to be viewed by the public. Deerhide stretched across a bowl of wood filled with sacred water, the Water Drum "signifies the heartbeat of the water, of the universe, of creation and of the people." The wooden bowl gives honor to the plants, the deerhide honors the animals, and the water within the life of Mother Earth. The drum is bound with a hoop that represents the circle in which all things move; birth, growth, and death, the circle of the seasons, the circle of our years.

There is no ecosystem on earth where mosses achieve greater prominence than in a *Sphagnum* bog. There is more living carbon in *Sphagnum* moss than in any other single genus on the planet. In terrestrial

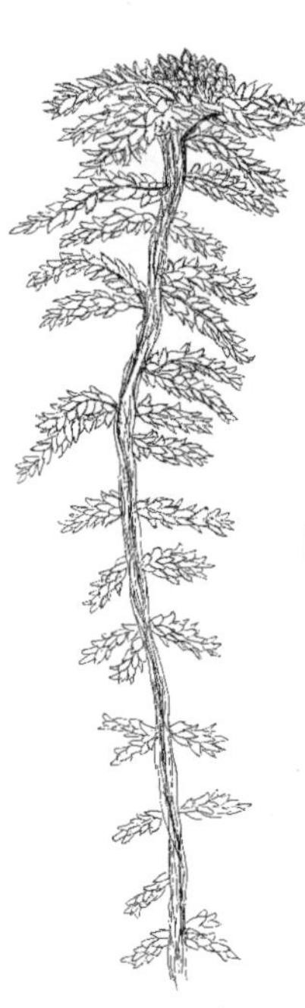
Individual Sphagnum plant

habitats, mosses are overshadowed by the vascular plants and assume relatively minor roles. But in bogs they are supreme. *Sphagnum* or peat mosses not only flourish in bogs, they create them. The acidic, waterlogged habitat is hostile to most higher plants. I know of no plant, large or small, which has the ability to engineer the physical environment more thoroughly than *Sphagnum* through the remarkable properties of the plant itself.

Every bit of ground in a bog is blanketed in *Sphagnum*. Actually, it's not ground at all. It's only water, cleverly held by the architecture of moss. I am walking on water, on a mat of *Sphagnum* moss that lies over the surface of the pond. Some of the pool is still visible at the center of the bog, a flat dark surface. Bog ponds are unusually still and glassy. The dark water draws your eye downward, in search of the unseen. No current disrupts the reflections of summer clouds, since the only source of water is rainfall. No stream runs in or out of this island of *Sphagnum*. The water is clear, stained the color of root beer with humic and tannic acids released from the slow decay of *Sphagnum*.

An individual stem of *Sphagnum* is reminiscent of an English sheepdog after a swim in the pond, dripping puddles onto the floor. *Sphagnum* has a great moplike head, the *capitulum*, held above the water. The rest of the plant is concealed by long, pendant branches that hang from the nodes of the stem. The leaves are tiny, just a thin membrane of green and they cling like sodden fish scales to the branches. If the mat is disturbed, *Sphagnum* even smells like a wet dog as sulfurous vapors are released from the muck below.

What amazes me most about *Sphagnum* is that most of the plant is dead. Under the microscope you see that every leaf has narrow bands of living cells which border the patches of dead cells, like green hedgerows around empty pastures. Only one cell in twenty is actually alive. The others are merely dead cell walls, skeletons surrounding the open space

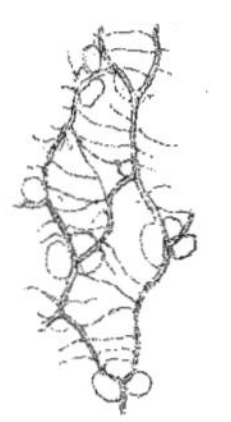

Porous cells of Sphagnum

where the cell contents used to be. These cells aren't diseased; they achieve their mature, fully functioning status only when they are dead. The walls of the cells are porous, peppered with tiny pores like a microscopic sieve. These perforated cells can't photosynthesize or reproduce and yet they are integral to the success of the plant. Their sole function is to hold water, lots of water. If you grab some *Sphagnum* from the seemingly solid surface of the bog, it comes up dripping. You can wring nearly a quart of water from a big handful of *Sphagnum*.

By allowing the dead cells to fill, *Sphagnum* can absorb as much as twenty times its weight in water. Its tremendous water-holding capacity allows *Sphagnum* to modify the ecosystem for its own purpose. The presence of *Sphagnum* causes the soil to become saturated, filling the spaces between soil particles that might otherwise hold air. Roots need to breathe, too, and the waterlogged peat creates an anaerobic rooting environment which most plants can't tolerate. This impedes the growth of trees, leaving the bog sunny and open.

The lack of oxygen in the sodden mat below the living *Sphagnum* also slows the growth of microbes. As a result, decomposition of the dead *Sphagnum* is extremely slow; it may persist relatively unaltered for centuries. The buried portions of the *Sphagnum* plant simply remain, year after year after year, gradually accumulating, filling the pond. My red sneaker, if I could find it in the depths of the bog, would not have decayed at all. Odd to think that a sneaker would outlast a person. In a hundred years, it may end up being the most tangible sign of my brief presence on the planet. I'm glad it was red.

This preservative effect yielded a stunning find for peat cutters who unearthed perfectly preserved bodies from a peat bog in Denmark two thousand years after burial. Archeological studies revealed these were the remains of Tollund people, Iron Age villagers known as the Bog People. Burial in the bog was no accident. Evidence suggests that these people were sacrificed in agricultural rituals, life offered up in exchange for a bountiful harvest. Their faces are surprisingly serene and their presence speaks the understanding that life is renewed only through death.

A side effect of the slow decomposition is that the minerals bound up in living things are not easily recycled in a bog. They persist in the peat as complex organic molecules that most plants can't absorb. This leads to severe nutrient deficiencies that exclude many vascular plants which can't tolerate the infertility. Most of the trees which do manage to take root in a bog are yellow and stunted, as a result. Nitrogen is in especially short supply, but some bog plants have evolved special adaptations to deal with this limitation: eating bugs.

Bogs are the exclusive homes for insectivorous plants like sundews, pitcher plants, and Venus' fly traps which perch on the *Sphagnum* mat. Deer flies and mosquitoes are plentiful in a bog and each one is a flying packet of concentrated animal nitrogen. The sticky traps and elaborate pitchers evolved to capture this nitrogen for the plant, gaining by carnivorous leaves what its roots could not provide.

Sphagnum is thorough in its manipulation of its environment. Not only does it produce waterlogged, nutrient-poor conditions; it also changes the pH, just for good measure. *Sphagnum* acidifies the water in which it grows, making it inhospitable for other plants. Release of acids allows *Sphagnum* to absorb scarce nutrients for its own use. The water at the edge of a bog may have a pH of 4.3, equivalent to dilute vinegar.

The acidity contributes to the antimicrobial properties of the moss. Most bacteria are inhibited by low pH. Because of this and its unsurpassed absorptive abilities, *Sphagnum* was once widely used as bandages. In World War I, when cotton supplies were limited by warfare in Egypt, sterile *Sphagnum* became the most widely used wound dressing in military hospitals.

The asymmetric ratio of 1:20 between living and dead cells in a single plant is mirrored in the structure of the entire bog. Most of the bog is dead, unseen. A *Sphagnum* bog is made of two levels, the deep dead peat and the thin surface of living moss. Only the top few inches of a *Sphagnum* plant are alive. The sunlit green capitulum and this year's branches are just the tip of a long, long column of *Sphagnum* tissue which may stretch several meters into the bog. Each year the living layer grows upward, farther and farther from the waters

Individual leaf of Sphagnum

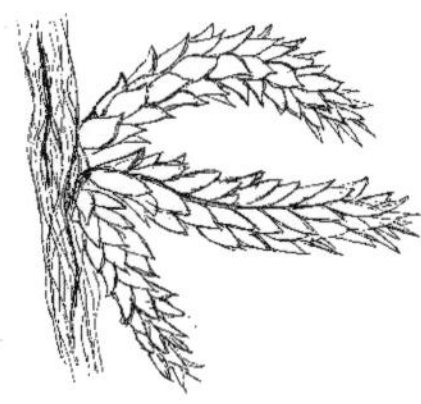

Clustered branches of Sphagnum

below. But the pendant branches droop downward, the dead cells wicking water up from the depths, carrying it to the living layer above.

Beneath this is the peat, the partially decomposed remains of *Sphagnum* plants that used to be on the living surface. The dead mosses are compressed by the weight of the water and the plants above, forcing them downward into the depths. This is the foundation of the bog, a huge sponge, holding water and steadily passing it upward, from the unseen to the seen.

Peat has a long history of human use, ranging from therapeutic baths in ancient Greece to ethanol generation today. Burning bricks of dry peat was an important heating source for many northern peoples. It is the smoke of slowly smoldering peat permeating malted grains that gives scotch whiskey its rich taste of autumn. The distinctive flavors of single malt scotches are said to originate in the qualities of the local peat cut from a particular moor. Peatlands are drained all over the world for growing certain specialty vegetable crops, like lettuce and onions.

The major commercial use is as a soil additive for gardens. I once had a garden at the edge of a floodplain, and the soil was so full of clay, we could have opened a pottery shop. I bought bales of peat to till into the soil. The fragments of organic matter help to keep the clay particles apart from one another, lightening the soil. People also dig it into their gardens to increase the water-holding capacity of a garden soil by the absorptive nature of all those dead cells. It can also be a sponge for nutrients, promoting their slow release to plants. Opening a bag of peat releases the scent of the bog. When I crumble the peat between my fingers, I remember its story, where it came from. The dry brown fibers that now see the light of day spent centuries beneath the dark water of the bog. And before that, they had a brief existence at the green surface, where dragonflies dove after mosquitoes snatching them away from the sundews. Commercial peat is mined from bogs that have been drained, some by nature, most by design. My garden and I are complicit in that enterprise and it troubles me. I prefer a bog wet and squishy between my toes.

Going barefoot is the best way to know a bog. Your feet will tell you things your eyes cannot. At first, the pillow-soft surface of the bog seems homogeneous, but as you walk through it, its complex pattern becomes apparent. There may be as many as fifteen different species of *Sphagnum* here, each with subtle variations in appearance and ecology. You don't really walk through a bog; it's more like barely controlled lurching. Your feet tentatively probe each spot, testing whether it will hold you before committing to a step, lest you join the bog people in becoming an artifact of history.

Kettle hole bogs are arranged in concentric circles of vegetation, rings of increasing age from the youngest edge of the open water to the high hummocks beneath the old tamaracks. The pattern is the product of the passage of time and *Sphagnum*'s power to transform the environment. Right at the pond's edge, at the youngest part of the bog, are *Sphagnum* species that live no place else, nearly submerged in the highly acidic water. The appearance of a solid mat is only an illusion. It is floating, dangling from the edge and even the weight of a bullfrog will push it under.

If you back carefully away from the edge, the mat becomes thicker, grown dense with the accumulating layers of moss. In the summer sun, it feels like a warm sponge beneath your feet. As you sink deeper, your toes curl around the submerged roots of bog-loving shrubs, running under the *Sphagnum* mat like taut wire springs beneath a soft mattress. Out here on the open mat, the *Sphagnum* rests on this framework. There are certain species of *Sphagnum* which inhabit only this zone of the bog, which is not usually submerged and therefore a bit less acidic. As the shrub roots continue to grow out toward the pond, these mat species will follow it, eventually closing in and hiding the open water under a *Sphagnum* blanket.

The next ring in the concentric circles of vegetation is the hummock zone. It's not quite so bouncy, since the depth of peat accumulation is greater in this older part of the bog. Here the difficulty in walking comes not just from the risk of sinking, but from uneven terrain. The bog surface is marked with hummocks of thick vegetation and thin spots in between. This is the spot where you might wish you had your shoes. The

Sphagnum mat is mingled with dead branches of shrubs, hidden beneath the soft moss surface and waiting to put you in line for a tetanus shot. Hummocks are formed by the interaction between the *Sphagnum* and the shrubs as each struggles for primacy. Just like your feet, the shrubs start to sink into the mat under their own weight. The surrounding blanket of *Sphagnum* sends shoots into the lower branches of the shrub, and wicks up water from below. This makes the shrub even heavier and it sinks some more, burying those branches. The cycle continues, the shrub growing upward, and the *Sphagnum* weighing it down. Eventually a cone-shaped hummock of shrub and moss builds up above the surface of the bog mat, as high as eighteen inches. In many cases the shrub dies, but its branches remain, entombed in the hummock.

At the small scale of mosses, the hummock presents a range of microclimates, analogous to the elevational zones of a high mountain. The bottom of the hummock is submerged in the mat, acid and soggy, while the top has become isolated from the water. The nature of *Sphagnum* allows water to be wicked up to the summit of the hummock. Nonetheless, the top is considerably drier, and less acidic. Not surprisingly, there are different species of *Sphagnum*, which inhabit each of these zones, creating a layer cake of mosses, each one adapted to the specific microclimate of the hummock slope, from the valley floor to the alpine summit. The presence of these small microclimates and the species of *Sphagnum* adapted to them are what contribute to the high biodiversity of bogs.

If you lay a hand on the hummock top, it's warm and dry on a summer day. Let your fingers probe downward and it gets progressively cooler and wetter. It's possible to reach your arm all the way through the hummock and into the peat below. It may be as much as fifty degrees cooler than the surface, since the dead air spaces in dry moss make an excellent insulator. The cold temperatures also slow decomposition. Traditionally, people who lived in the boggy taiga would utilize the cold peat as a refrigerator for freshly killed game. When I was an undergraduate, one of my professors, Ed Ketchledge, used to torment we students with this phenomenon. We'd be on a bog field trip, struggling in the heat and swatting deerflies, sipping from our tepid canteens. He'd calmly step over to a certain hummock, reach down inside and pull out a cold beer that he had stashed there on his last visit. That's a lesson not soon forgotten.

The hummock tops are often dry enough that *Sphagnum* can no longer live there, and other mosses will colonize. These high hummocks provide the only spots where trees could get established, with their roots up above the saturated peat. You see little seedlings of tamaracks and spruces perched on top. A few of them actually succeed, and begin to form an open bog forest. Another group of *Sphagnum* species flourishes under these trees, where the peat is deep and solid.

In these thick peat deposits, paleoecologists can read the history of the land. They slide a long shining cylinder into the bog, cutting through layers of undecomposed plants, and extract a core of peat. By the plants that are present, the pollen grains trapped there, and the chemistry of the organic matter they can discern the changes in the land. Changes in vegetation, changes in the climate, stretching thousands of years before, are all recorded there. What will they read in the layer that represents our time, our evanescent moment at the surface? We are responsible for that.

I love listening to a bog, the papery rustle of dragonfly wings, the banjo twang of a green frog, the occasional hiss of sedges moving in the breeze. On a hot summer day, if you're very quiet, you can witness the smallest discernable sound I know—the "pop" of *Sphagnum* capsules. It's hard to imagine that a sound emitted by a capsule only one millimeter long could be audible. Their capsules, tiny urns on a short stalk above the moss, explode like a popgun. The heat of the sun builds up air pressure inside the capsule, until the top blows off, propelling the spores upward. In the stillness, listening intently, I thought I heard the sound of the Water Drum.

A quaking bog feels to me like the living embodiment of the Water Drum, the mat of *Sphagnum*, stretched across the surface of the water, held in a granite bowl carved by a glacier. The *Sphagnum* is the living membrane stretched between two shores, creating a meeting place for earth and sky, embracing the water within. I am standing quietly on the surface of an earthly Drum, my feet supported by the floating *Sphagnum*, responding to the smallest movement, rippling under my shifting weight. I start to dance. In the old way, heel and toe, in slow tempo, each footfall rippling across the bog and answered by the returning wave rising to meet my step. My feet make a drumbeat on the surface and the whole bog is set in rhythmic motion.

The soft peat below responds to my step, compressing with the downbeat and springing back. It too is dancing deep beneath me, sending its energy up to the surface. Dancing on the *Sphagnum*, buoyant on the surface of the peat, I feel the power of connection with what has come before, the deep peat of memory holding me up. The drumbeat of my feet calls up echoes from the deepest peat, the oldest time. The pulsing rhythm, persistent, wakens the old ones and as I dance I can hear their faraway songs, the songs of the Water Drum in the medicine lodge, the songs they made winnowing wild rice on the shores of a vast blue lake, mingled with the voices of loons. Emerging like a vapor from the deep peat of memory come the farewell songs and the cries of the people marched off their beloved homelands, prodded down the Trail of Death at the point of a bayonet that brought them to the drylands of Oklahoma where no loons sang. Up, up through the peat, up through time their voices are rising, voices of the good sisters of St. Mary's teaching the red children their duplicitous catechism.

Dancing, sending the message of my presence through the peat, I can feel in response the rumbling vibration of the train wheels rolling eastward, carrying my grandfather, just nine years old, to Carlisle Indian School, where they dance to the persistent rhythm of "Kill the Indian to Save the Man." Dark peat, dark times, when the Water Drum nearly lost its voice. Memory, like peat, connects the long dead and the living. Spirit, like water, was wicked up from below, hand to hand from the watery depths to the parched surface where my grandfather lived in boarding-school barracks, sustaining him. They did not Kill the Indian. For today, I am dancing, on a Water Drum of peat in a country of vast blue lakes where loons are calling. Dancing, my feet sending the message of my presence in waves through the peat, and in waves of memory, they send back the message of their presence. We are still here. Like the living surface of the *Sphagnum*, the sunlit green layer at the top of a column of dark accumulated peat, individually ephemeral, collectively enduring. We are still here.

Maybe my presence need not be marked by more than my red sneaker. Just by continuing, I honor the lives of my ancestors and form the foundation for my grandchildren. We are profoundly responsible for one another. When we gather and dance in the elder's footsteps, we honor that link. When we steward the earth for our children, we are living like *Sphagnum*.

Portrait of Splachnum

The jet stream flows through the stratosphere like a muddy river. It cuts from one shore and deposits on the other, homogenizing its load of sediment. Swept along in the current are airborne seeds and spores, keeping company with vagrant spiders. Every continent is awash in the same aerial plankton. The wonder is not that the earth should be so richly populated, but that it is not everywhere the same. Somehow, each wandering spore finds its way home.

This global cloud of spores powders every surface with the possibility of mosses. I saw the same species in my driveway in upstate New York that I encountered the next morning in the sidewalk cracks of Caracas. This same species chinks cracks in the cinder blocks of Antarctic outposts. It is not the proximity to the equator that matters, but the singular chemistry of pavement which makes a home.

The boundaries that define where a particular species of moss calls home are often more narrow. Some are strictly aquatic, some are terrestrial. Epiphytic mosses restrict themselves to the branches of trees, but some epiphytes settle only on sugar maples and others only in the rotten knotholes of sugar maples growing on limestone. There are generalists that are ubiquitous on any patch of open soil, and specialists which favor the dirt tossed up by the diggings of pocket gophers in the tall-grass prairie. Some saxicolous (rock-dwelling) mosses can live on granite, others only on limestone and *Mielichoferia* only on rocks containing copper.

But no moss is more fastidious in its choice of habitats than *Splachnum*. Absent from the usual mossy haunts, *Splachnum* is found only in bogs. Not among the commoners like *Sphagnum* that build the peaty hummocks, not along the margin of the blackwater pools. *Splachnum ampullulaceum* occurs in one, and only one, place in the bog. On deer droppings. On white-tailed deer droppings. On white-tailed deer droppings which have lain on the peat for four weeks. In July.

I've never found *Splachnum* by looking for it. Days before my moss class was scheduled to begin, I would go to a quaking bog in the heart of the Adirondacks in hopes of finding a patch to show my students. I'd found it there before, but only when I was looking for something else. Squishing through the muck my footsteps would release the faintly sulfurous gases. I searched over the carpet of peat, finding pockets of rare pitcher plants, sundews, and spider webs stretched over the branches of bog laurel. I found plenty of deer droppings, and coyote scat too, but the tidy piles of brown pellets are empty.

leaf of Splachnum

While all are very rare, any bog could have as many as three different species of *Splachnum* residing there. *Splachnum ampullulaceum* inhabits the droppings of white-tailed deer. Had a wolf or coyote followed the scent of the deer into the bog, its droppings would have been colonized by another species, *S. luteum*. The chemistry of carnivore dung is sufficiently distinct from that of herbivores to support a different species. If a moose strode through the bog, and made a contribution to the local nitrogen economy, its droppings would be of little use to either of the other *Splachnum* species. The moose droppings have their own loyal follower.

The family to which *Splachnum* belongs includes several other mosses with an affinity for animal nitrogen. *Tetraplodon* and *Tayloria* can be found on humus, but primarily inhabit animal remains such as bones and owl pellets. I once found an elk skull lying beneath a stand of pines, with the jawbone tufted with *Tetraplodon*.

The set of circumstances that converge to bring *Splachnum* into the world is highly improbable. Ripening cranberries draw the doe to the bog. She stands and grazes with ears alert, flirting with the risk of coyotes. Minutes after she has paused, the droppings continue to steam. Her hoof prints leave indentations in the peat, which fill with water and leave a trail of tiny ponds behind her. The droppings send out an invitation written in wafting molecules of ammonia and butyric acid. Beetles and bees are oblivious to this signal, and go on about their work. But all over the bog, flies give up their meandering flights and antennae quiver in recognition. Flies cluster on the fresh droppings and lap up the salty fluids that are beginning to crystallize on the surface of the pellets. Gravid females probe the dung and insert glistening white eggs down into the warmth. Their bristles leave behind traces from their earlier foraging trips among the day's dung, delivering spores of *Splachnum* on their footprints.

The spores germinate quickly in the wet droppings and trap the pellets in a net of green threads. Speed is of the essence. Growth must outpace the decay of the droppings, or else the moss will find itself without a home beneath its feet. The nutrients in the dung speed development, so that in just a few weeks the dung is concealed beneath a lawn of pure *Splachnum*. Like all other plants, mosses are confronted with choices about the allocation of their energies to growth or to reproduction. Investment in long-lived stems and leaves can pay large dividends in the future, allowing the plant to elbow aside competitors and maintain its dominance in the community. Reproduction is then delayed, while the limited energy supply is used to fuel growth. This strategy is effective in stable habitats, where the opportunities for reproduction stretch into the future. The habitat is likely to outlast the moss. But where the habitat is transient, the plant benefits most by investing its energy in mobility. Getting stuck in a disappearing habitat is a sure means of local extinction. The plant must quickly produce a crop of airborne spores, to disperse itself to new habitats, before the old one deteriorates. *Splachnum* is a just such a fugitive species, quickly colonizing one dung pile and then fleeing to the next when the droppings rot away.

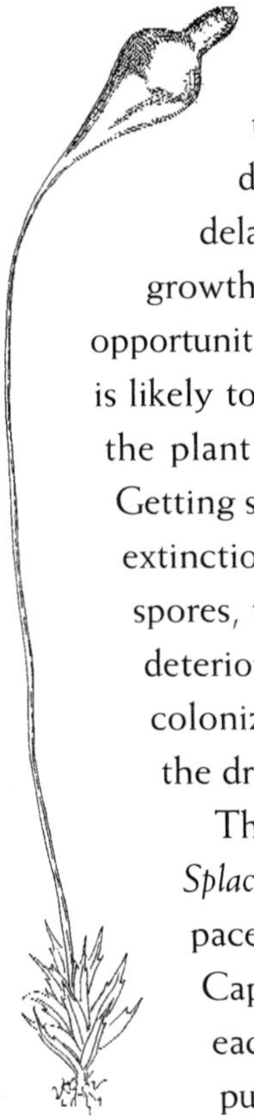

The urgency for departure pulses through a developing *Splachnum* colony. With remarkable speed among the snail-paced mosses, the sporophytes seem to appear overnight. Capsules burgeoning with spores rise above the leaves, each pushed upward on an elevated stalk. No other moss puts on such a gaudy display of unbridled reproduction. In unmossly shades of pink and yellow, the capsules

Splachnum

tower over the leaves and wave in the breeze. The capsules swell to bursting and exude a sticky mass of colored spores. More modest mosses rely on the wind to carry their offspring, and the wind needs no extravagances to attract it. Since *Splachnum* can grow only on droppings, and nowhere else, the wind cannot be trusted with dispersal. Escape of the spores is successful only if they have both a means of travel and a reserved ticket for a particular destination. In the monotonous green of the bog, flies are attracted to the cotton candy colors of *Splachnum*, mistaking them for flowers. Rooting about in the moss for non-existent nectar the flies become coated with the sticky spores. When the scent of fresh deer droppings arrives on the breeze, the flies seek it out and leave *Splachnum*-coated footprints in the steaming dung. And so, on some dewy morning when I am picking blueberries in the bog, a *Splachnum* bouquet will appear, unbidden, at my feet.

The Owner

The letter had no return address. I'd been summoned by an invisible man, with an offer I couldn't refuse. The letter, on thick white paper, requested my "expert services as a bryologist, to consult on an ecosystem restoration project". It sounded pretty good.

The goal was to "create an exact replica of the flora of the Appalachians, in a native plant garden." The owner was "committed to authenticity and wished to ensure that mosses were included in the restoration." Not only that, he requested "guidance on matching the correct species of moss to the proper rock types in the landscape." That was to be my task should I accept their generous offer. The letter had no personal signature, just the name of the garden. I read the letter again. It sounded too good to be true. There are few people interested in ecological restoration, let alone restoration of mosses. One of my research interests at the time was to understand how mosses were able to establish themselves on bare rock. This invitation seemed like the perfect match. I was intrigued by the project, and as a new professor, I was admittedly flattered by the prospect of using my expertise and getting paid consultant fees to do it. The letter had an air of urgency about it, so I made plans to go as soon as possible.

I pulled over to the roadside to unfold the directions which lay on the seat beside me. My instructions requested precise punctuality and I was trying to oblige. I'd been driving since dawn to reach this lovely valley, where bluebirds dove across the winding road into impossibly green pastures of June. An old rock wall ran beside the road and even from the car I could admire the mossy cover it had amassed over its long life. Down South, they call these "slave fences" in acknowledgement of the hands that laid the rocks. A century's worth of *Brachythecium* softens the edges and the memory. The directions had me follow the stone wall until the chain-link fence began. "Turn left to the gate. It will open at

10:00 a.m." Indeed, just as I arrived, the massive gate rolled smoothly away to the side, responding to some unseen commander. It was startling to find such security in this valley, which seemed more suited to horse-drawn wagons than electric eyes.

I started up the steep hill, gravel crunching under my tires. I had four minutes. Around a bend in the road, I could see a rooster tail of dust against the blue morning sky. It was creeping up the hill ahead of me so slowly I know I'd be late. Around the switchback, laboring up the hill, I got a glimpse of what I was following. My brain rejected the image. Trees don't move. But there it was again—the bare spring branches of a tree visible against the hillside, and it was heading uphill. I could see clearly now. It was an oak tree, riding piggyback on a flatbed truck. Now this was not a standard nursery-size tree with tidy burlapped root ball. No, this was a big old granddaddy oak. We had one like it on our farm in Kentucky, a huge burr oak with low spreading branches that cast a pool of shade the size of a house. It took two of us to reach our hands around it. There is no way you can move trees of this stature. And yet here it was—strapped to a truck like a circus elephant on a parade float. The root ball was twenty feet across and tethered to the truck with steel cables. The truck pulled over and steam rose from under the hood as I passed by, staring.

The road ended in a lot full of construction vehicles, all with engines running. The scraped earth was surrounded by a collection of barns and open-doored garages. I parked next to a row of dusty Jeeps and looked around for my host. There were dozens of people moving about, at a frenetic pace that reminded me of an anthill disturbed. Trucks were loaded and sped away. Most of the workers were dark and small. They wore blue jumpsuits and called to each other in Spanish. One man stood out in his red shirt and white hardhat. His folded arms announced that he was waiting for me and that I was late.

Introductions were brief. Looking at his watch, he commented that the owner monitors the use of consultants' time carefully. Time is money. He pulled the radio from his belt to inform some higher authority of my arrival. I was handed off to a young man who emerged from an office in the barn. His shy smile and warm handshake felt like an apology for the brusque welcome and he seemed eager to escort me away from the center of activity. This was Matt, fresh out of college with a newly minted horticulture diploma. It was his second year working on the garden and it was he who had petitioned the owner to invite in a moss consultant to help with his assigned and somewhat overwhelming task of moss restoration. Matt knew that his work had high visibility in the design of the garden. Apparently the owner was especially fond of mosses, so the pressure for success was high. His goal was botanical accuracy in the plantings, and erasing the newness of the garden by introducing mosses throughout the landscape. Matt strode ahead and I followed, down a newly poured sidewalk through the construction site. He wanted me to look first at the moss garden. We could cut through the house to get there, since the owner was not at home.

The brand-new house had the look of an old manor and was surrounded by huge trees set into the bare soil: tulip poplar, buckeye, and a gnarled sycamore. Each was anchored by guy wires and black tubing ran through the canopy. The oak I had met on the road had arrived and a gaping hole stood ready for its roots. It would stand right outside a wall of leaded glass windows. "I didn't know you could buy trees that big," I said. "You can't," Matt replied. "We have to buy the land and then dig them up. We have the biggest tree spade in the world." He watched the shock on my face for a moment and then looked away, picking sheepishly at his hands, and then recovered his professional demeanor. "This one came in from Kentucky." He explained that every tree was treated with chemicals to reduce transplant shock and then a drip irrigation system was installed in its canopy. It operated on a timer and delivered a spray of nutrients and hormones to stimulate root growth. The garden had a team of expert arborists, and they had not lost a tree yet. The entire grove around the house had been transplanted, trees cut from their soil by enormous tree spades and trucked here to restore an ecosystem.

Matt disarmed the security system with a swipe of a card and we passed into the air-conditioned dimness of the house. This side entryway was a virtual gallery of African art. Carved masks and geometric weavings lined the walls. A cowhide drum, a wooden flute stood on stone pedestals and I stopped and stared. "It's all authentic," Matt told me, proudly. "He's a collector." He stood by as I looked around and let my amazement polish his status. Each piece was labeled with its village of origin and the name of the artist. It was an impressive display. At the center of the atrium in a discreetly alarmed case spotlights were focused on an elaborate hairpiece. Its intricate design of bees and flowers was carved from luminous ivory. I was immediately struck by how out of place it seemed on its velvet platform, more like a stolen treasure than a work of art. How much more beautiful it would have been in the black oiled hair of the artist's wife. And more authentic. In a display case, a thing becomes only a facsimile of itself, like the drum hung on the gallery wall. A drum becomes authentic when human hand meets wood and hide. Only then do they fulfill its intention.

We passed through a vaulted room holding the swimming pool and I was unquestionably dazzled. The pool stood in a room decorated with hand-painted tiles and lush tropical vegetation. The marble floor glowed and the pool gurgled invitingly. I felt as if I had walked onto a movie set. Lounge chairs were casually placed around the pool, thick towels folded and ready for the convenience of guests. The stemware

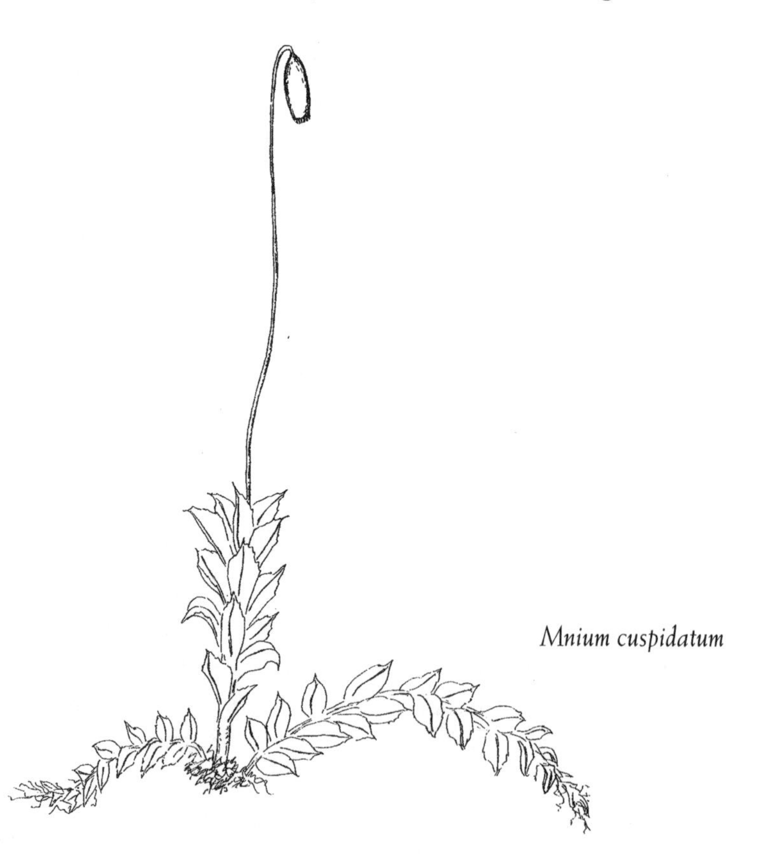

Mnium cuspidatum

arranged on the patio tables was exactly the same ruby color as the towels. "The owner will be here this weekend," Matt said, waving his hand at the preparations. We finally made it to the kitchen, where I was offered a drink of water from a Dixie cup.

The courtyard garden at the center of the house was Matt's first concern. He walked a little taller through the lush green he'd created. Every manner of tropical plant was there: birds of paradise, orchids, tree ferns. Paving stones made a path completely carpeted by *Mnium*. It was a spectacular array of feathery green as smooth as the mossy lawn of a Japanese garden. He'd been having trouble keeping the moss alive and constantly had to go back out to the woods to get more to maintain the unbroken surface. So we spoke of water chemistry and soil conditions as he scribbled in his notebook. I was feeling useful at last, giving advice on matching the right moss species to the garden environment so that it might naturally regenerate itself. I cautioned him about the ethics of wild gathering. The woods shouldn't be a nursery for his garden. His garden would be successful only when it became self sustaining. At the center of the garden stood a sculptured rock taller than either of us and beautifully covered with mosses. Each carefully chosen clump accented the irregularities of the boulder. An eroded pocket in the rock was filled with a perfect circlet of *Bryum*. The artistry rivaled any piece we had seen in the gallery and yet it struck the wrong note; the collection was only an illusion of nature. *Plagiothecium* can't grow in crevices like that, and *Racomitrium* wouldn't share a habitat with *Anomodon*, despite the beauty of their colors side by side. I wondered how this beautiful but synthetic creation passed the owner's standard of authenticity. The mosses had been reduced from living things to mere art materials, ill used. "How did you get these to grow like this?" I asked. "It's very—unusual," I hedged. Matt smiled like a kid who had outsmarted his teacher and answered, "Superglue."

Moss gardens are a demanding exercise and I was impressed by what they'd accomplished. But where was the ecosystem restoration that all those trucks and workers were devoting themselves to? When at last we went outside there was no native plant garden, just the bare skeleton of a golf course under construction. A little tornado of dust rose up from the bare ground. The cart paths were lined with great slabs of rock placed in anticipation of the grass to come. The rocks themselves were beautiful behemoths of mica schist, the native bedrock which glittered like gold in the spring sunshine. A drainage pond had been cut into the golf course and was bordered by a wall from which the rocks had been recently quarried in terraced steps.

Matt led me up to the top of the quarry wall and we looked out over the scene. Bulldozers scraped and shoved, reshaping the land for the play of the game. Matt explained that the owner didn't like to see the raw rock around the pond. It looked as if it had just been blasted, which in fact, of course, it had. The owner requested that I give them a method for growing mosses to carpet the quarry wall. "It's the backdrop for the golf course and the owner wants it to look like it's been here for years," explained Matt. "Like an old English course. The mosses will make it look old, so we need to get them growing". It's way too big to use Superglue.

There are only a handful of moss species that can colonize the harsh surface of acidic rocks and none of them is exactly luxuriant. Most form brittle, blackish crusts that are well adapted to withstand the stressful environment, but which would not even be noticed by a passing golfer. The black color of mosses which grow in full sun is produced by anthocyanin pigments which shield the plant from the damaging ultraviolet wavelengths that their shade-loving counterparts can avoid. I explained that moss growth depends heavily on a supply of water that this bare rock wall simply didn't possess. Without moisture, even centuries of moss growth would yield only a black crust. "Oh, that's not a problem," Matt responded. "We can install a misting system. We could run a waterfall over the whole thing, if that would help." Evidently, money was no object. But it was not money that the rocks required, it was time. And the "time is money" equation doesn't work in reverse.

I tried to be diplomatic in my answer. Even with a watering system, the green carpets envisioned by the owner would take generations to grow. In fact, growth itself was not the issue. The critical step in moss growth comes at the birth of a colony. I'd spent considerable effort in researching how mosses decide to take hold of a rock. We have an idea of "how," but the "why" is very poorly understood. Wind-blown spores, finer than powder, must be stimulated to germinate by just the right microclimatic conditions. Raw rock is inhospitable to mosses. The surface must first be weathered by wind and water, and then etched by the acids produced by a lichen crust. The spore then forms delicate green filaments called *protonema*, which attach firmly to the rock. If they survive, tiny buds will form and sprout into leafy shoots. In experiment after experiment, we see that the probability of a single spore ever producing a single moss shoot is vanishingly small. And yet under the right circumstances, and given enough time, mosses will blanket a rock, like the old stone slave fence. So creating a colony of moss on a rock is no small feat, a mysterious and singular phenomenon which I have no idea of how to reproduce. Much as I would have liked to be the successful problem-solving consultant, I had to deliver the bad news. It couldn't be done.

Each time we changed locations, Matt would call in on the radio. I wondered who would care exactly where we were. We walked back up to the house where truckloads of huge rocks were unloading. "This is where they'll build the terrace," said Matt. "The owner wants mosses to grow on all these rocks, too. It will all be shaded, so do you think we could grow them here? With a mist system installed?" Matt pressed his case. If giant oaks could be transplanted, why not mosses? Couldn't you get mosses on the rocks simply by transplanting them? If you could provide the right shade and water and temperature, shouldn't they survive? Again, the answer was not what the owner would want to hear.

You'd think that, without the complication of having roots, moving mosses to a new home would be quite simple. But mosses are not like the plants in my perennial bed that I can move about the garden like rearranging furniture. A few soil-dwelling mosses like *Polytrichum* can be transported much like a grass sod, but rock-loving mosses are inordinately resistant to domestication. Even with the greatest care, moss transplants from rock to rock are generally doomed for failure. It may be that dislodging them tears nearly invisible rhizoids or crushes cells beyond repair. Or our studied duplication of their habitat is lacking in some key ingredient. We don't really know. But they nearly always die. I wonder if it's a kind of homesickness. Mosses have an intense bond to their places that few contemporary humans can understand. They must be born in a place to flourish there. Their lives are supported by the influences of previous generations of lichens and mosses, who made the rock into home. In that initial settling of spores they make their choice and stick to it. Relocation is not for them.

"Well, what about sowing them, then?" Matt asked. He looked so hopeful. This was his first job, with a demanding boss and a nearly impossible task. I felt compelled to meet his hope halfway and, perhaps, to salvage his expectations of my supposed expertise.

There's no good science to suggest how to start mosses on rocks, but there are reports of a kind of moss magic in the folklore of gardeners. It's worth a try, I suppose. Garden enthusiasts have long looked for ways to speed up the growth of mosses on rock walls, to induce bare rock to assume the patina of mossy old age. I've heard of dousing the wall repeatedly with acid. Supposedly, it dissolves the surface of the rock and creates tiny pores where mosses can gain a foothold. In a way it mimics the action of lichen acid, slowly eating away at the rock. Other gardeners swear by swabbing a slurry of horse manure over the rocks. It's a bit rank at first, but the mosses seem to follow in short order. The most common recommendation is much more hygienic, a moss milkshake. The recipe goes like this: collect the moss species of interest from similar rocks in the forest. Be sure to take only mosses growing in the same conditions that are found in your garden. The same kind of rock, the same light and humidity. No cutting corners allowed, the mosses will know the difference. The moss is then placed in a blender with a quart of buttermilk and whirred to a green froth. Painting this mixture onto rocks is said to yield a coating of moss within a year or two. There's a great variety of recipes out there, some using yogurt, egg whites, brewers yeast, and other household items. Hypothetically, there's some sense to these concoctions. Mosses can indeed regenerate themselves from severed leaves and stems. Under the right conditions a fragment will put out protonema to anchor itself to the new substrate and tiny little shoots can arise in

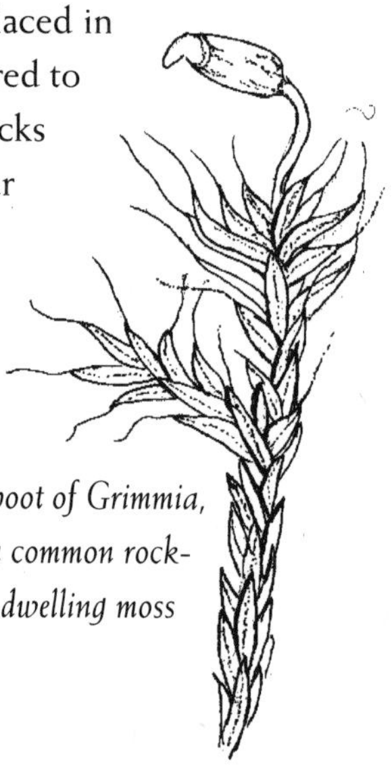

Shoot of Grimmia, a common rock-dwelling moss

this way. Mosses propagate themselves this way in nature, so perhaps a blender will help the process along. Many mosses prefer an acidic habitat, which might be provided by the buttermilk, at least until the first rain.

Since Matt was grasping at straws, I promised to write up the recipes for moss milkshakes, but cautioned that I had little confidence in any technique to grow instant moss.

We strolled along the site of the new terrace, just talking. There was a rocky flower bed along the path, full of native spring wildflowers. There were trillium and bellwort and a whole patch of leaves which I recognized as lady's slipper. Protected species, every one. Was this what they meant by an ecological restoration? A flower bed? I asked where the plants had come from and along with a "none-of-your-business," look I was assured that they'd been bought from a nursery which grew all their own plants. Indeed, each still had a nursery tag. Not a single one had been taken from the wild, he said with emphasis.

All day, Matt had made a concerted effort to adhere to a close-mouthed professionalism, but gradually his natural ease and openness couldn't be held back. He reminded me of my own students, eager to be out in the world and make a difference. This position was his first job offer and it had seemed too good to be true. The work was creative and he was better paid than he had thought possible for a rookie. After he'd been here for a year or so, he had had some doubts about the way things worked here and thought about moving on. But the owner had offered him a raise, if he'd stay. He'd just bought a nice little house and had a baby on the way, so he'd be around for awhile.

Back in view of the man in the white hardhat, Matt picked up his pace and strode across the construction yard like a man with a purpose, speaking into his radio. I followed along behind and hoped that I too presented the image of a busy professional. "Time is money," I heard him say in my head. We headed down one of the many roads that ran from the yard like spokes of a wheel.

When we were out of sight of the buildings, Matt looked once more over his shoulder and slowed down. "Do you mind if we walk cross-lots?" he asked. We stepped off the road into the trees and in only a few steps the smell of diesel was washed away by the smell of the spring woods.

In the concealment of the trees, he visibly relaxed. He grinned as he turned off his radio with a conspiratorial look and stuffed his hat into his back pocket. We suddenly felt like kids sneaking off from school to go fishing. "It's not too far," he said. "I want you to see what the native mosses are like here. Maybe you could tell if they're the right ones for trying to plant on the terrace. I'd like to try that milkshake method." He led me cross-country through the oak woods. In places, there were rocks strewn on the forest floor and I paused to look at their mosses. Matt was impatient. "We don't need to bother with these, the good stuff is right up here." He was right.

At the top of the rocky ridge, the land fell away sharply to a shady glen below. We scrambled down over the ledges of the massive outcrop, taking care not to scuff up the carpets of moss. The Appalachian bedrock here was folded and contorted by eons of geologic pressure, and then rearranged by the action of glaciers. The result was a fractured stone sculpture of unlikely angles, a cubist painting of a mossy landscape. The surface of each rock was etched by time into crevices like the wrinkles on an old man's face. Black trails of *Orthotrichum* traced the crevices, and thick beds of *Brachythecium* lay on the moist ledges. I thought I could see here Matt's inspiration for his superglued creation in the garden. It was lovely, a breathtaking tapestry of old mosses. Matt showed off every nook and cranny of the outcrop with evident familiarity. I suspect he'd played hooky and come here more than once. "This is exactly what the owner wants it to look like up at the terrace," he said. "I brought him here once and he just loved it. I just need to find a way to get it to grow like this up at the house." Somehow, I was getting the impression that I had not communicated the problem very well. One more time I launched my explanation of the relationship between time and mosses. The moss beds on this outcrop were probably centuries old. If it were possible to exactly duplicate this microclimate, and then sow these same species in a moss milkshake, you might have a chance. But, even then it would take years. Matt wrote it all down.

We walked back out to the road and checked our watches. Our appointed hours for consulting were over. Matt confided that the owner was extremely frugal and, especially with hiring outsiders, schedules had to be adhered to. The workers were climbing into trucks where they'd be driven down to the front gate, which was locked at 5 p.m. Standing by the cars, Matt gave me instructions from the owner about the report he would like to receive in three days' time. Before I left I had to ask, since no one ever said his name. "Who is the owner? Whose idea is behind this project?" His practiced answer came back quickly without his meeting my eyes. "I'm not at liberty to say. He's a very wealthy man." That much I had deduced.

Driving out to the gate, I scanned the landscape for signs of the ecosystem restoration I had never seen. But all I saw was the house and the golf course. It was too good to be true. I puzzled over the namelessness, the invisibility of this powerful man who had amassed all these resources to make a garden. Was this the anonymity of a discreet philanthropist or the concealment of a notorious identity?

My imminent departure was radioed ahead and when I reached the edge of the property, the gate opened and smoothly shut behind me.

꩜

Back at my office I wrote up an innocuous little report. I tried to educate The Owner as to the near impossibility of the task he proposed. All the money in the world won't get mosses to grow quickly on bare rock. That takes time. I included a species list of all the species we had seen, their environmental requirements, and guidelines for choosing the right species for a moss garden. Like a good academic, I suggested that if they really wanted to grow moss on rocks they should consider underwriting a cooperative research project. And I included the milkshake recipes for buttermilk and manure—who knows?

A few weeks later I got my check in the mail. I can't say that I felt very pleased with the work. What had been advertised as an educational project in vegetation restoration looked suspiciously like a tax shelter to landscape the new home of a wealthy man with a love for mosses and a passion for control. There may well have been good restoration work going on, off where the men in trucks were going, but I never saw it.

So I was surprised when Matt phoned me a year later. He asked if I could come down and help again. He said they'd made a lot of progress and he was eager to show me the garden. When I arrived he was nowhere to be seen. I was escorted off by a brisk young woman who had been delegated to show me the gardens. I asked about Matt, and she said he'd been reassigned to another project, maybe the azalea garden. She whisked me over to the house. "The Owner wanted you to see what they've done with the mosses on the terrace. We just finished last month."

What a transformation had been wrought. The place had aged a century in only twelve months. The Kentucky oak looked as if it had been born there and green lawns had materialized over the construction rubble. Where the pile of bare rocks had stood last spring, there was now a stunning replica of the native flora of an Appalachian ridgetop. Flame azaleas with sinuous trunks cast a light shade over the dark rocks, placed in a convincing jumble so that pitch pines appeared to emerge from the deep crevices. Clusters of bracken fern and myrica gale posed along the pathways which led to gatherings of weathered garden chairs. It did indeed look old. And much to my amazement, every rock was upholstered with mosses, lovely thick carpets of exactly the right species. *Brachythecium* capped the rocktops, with *Hedwigia* trailing down the sides. *Orthotrichum* artfully followed the etched ridges in the stone, like black calligraphy on old parchment. It was breathtaking. It was perfect in every detail. It was two weeks old. Perhaps I'd better rethink the merits of moss milkshakes, if this was the outcome.

My attendant was not much interested in my effusive praise of the garden. She was on a schedule and hurried me off to a patio on the other side of the house. It was a beautiful expanse of flagstone paving under the transplanted trees. "The Owner would like to know how to get rid of the moss that keeps coming up between the stones," she said, and waited with pen poised over notebook. I had no answer. All this work to get mosses to grow in one place, but where they spontaneously appeared he wanted them exterminated.

Back we went to the main staging area where earth-moving machinery rolled in and out. The squawking radios, uniformed men, and sense of urgency made it feel like a military operation. The hard hat sergeants rode around in jeeps while open trucks of Guatemalan infantrymen bearing shovels and pruning saws were carried away, all under the orders of the Owner.

I too was bustled into a Jeep and we bounced along a rough new road, cut like a gash into the oak woods. A driver had been sent for me, but offered little in the way of information as to where we were going. I wondered if I'd get to see Matt. The driver barked into his radio that we were nearly there. The makeshift road ended in a small clearing where a bright yellow crane stood. Empty pallets were stacked in the sun. In the shade of the perimeter were mysterious figures shrouded in burlap and wrapped with baling twine, like so many statues waiting to be unveiled. A cluster of muscled men stood in the woods, with their hard hats bowed together in consultation. One came forward and enthusiastically introduced himself as Peter, a designer specializing in natural rock. He was just ever so glad to see me because they needed advice before proceeding any further. He was here from Ireland and spoke with a lovely lilt. The Owner had brought him over especially for the job. He was concerned that they might be botching the moss bit and so would I come take a look? We joined the clustered men and they took their time sizing up the newcomer, the moss lady.

These gentlemen were introduced to me as the precision explosives team. They'd been sent for from Italy, a crack group of stonecutters. Before us stood the object of their scrutiny, a craggy rock outcrop covered deep in mosses. I recognized it immediately as the beautiful little glen where Matt had brought me the previous year. Half of it was gone. They were quite a dedicated team. Peter, the rock designer, would select the sectors of the cliff which were most beautiful, a place where a quartz vein ran through the schist and the mosses were especially well placed. The stonecutters would then carefully calculate the position for the precision charges and blow the rock off the cliff face. Lesser men then hoisted the rock with the crane and placed it on a pallet where it was wrapped in moist burlap to protect the precious mosses. I suddenly understood that the lovely rocks on the terrace owed nothing to applications of buttermilk. I felt my hands grow clammy.

They had so many questions. Should they burlap the rocks before the blast? Was Peter choosing correctly—would these particular mosses survive the move? How long could the mosses stay wrapped? Could I work with Peter to make recommendations about exactly where to place each rock for the well-being of the mosses? The Owner was upset that the mosses seemed to lose vigor once they were placed in the landscape. The cost of extracting each rock was very high and he didn't want to waste a single one. They considered me a new member of their team, hired to do this thing. I scanned all their faces for a sign of dissension but I saw only eagerness to get the job done. I felt numb, and somehow trapped as the Owner's hired gun. I had never dreamed that my advice would be put to such a use, that I would become an unwitting consultant for destruction.

The workers were very thorough in their care of the stolen mosses, and genuine in their concern for their well-being. They were watered and carefully swaddled in burlap for their removal. They'd do whatever I told them to help them survive. Once separated from their homes, the mosses seemed to sicken and the lush green turned to yellow. The Owner didn't want to waste money transplanting rock if the mosses were going to die. So, they had set up a triage area to try and nurse the likely candidates back to health and cull out the ones who wouldn't make it. The facility was a big white tent set up in a meadow alongside the road back to the house. It looked for all the world like a field hospital set up for the wounded. Shade screens were rolled down the sides to maintain humidity within. Mist nozzles sprayed out jets of water. No expense had been spared. Lying on their pallets were the blast-scarred rocks with their mosses lying sickly upon them.

My task was to make diagnoses and write the prescriptions. Which could safely be moved to the house and which should be abandoned? I thought of the doctors consigned to meet the slave ships at the shore. They would inspect the human cargo to pick out the healthiest for sale, the ones most likely to survive in their transplanted environment. And which was the lesser evil? To be sold into bondage or to be left behind to die? I wandered among the ailing rocks and felt as dislocated and powerless as they were. I wanted to shout at them to stop, but it was too late. And I was an accomplice to this. I don't remember what I said. I hope I said to save them all.

I wanted to see the Owner and confront him face to face with this betrayal, but he was invisible. Who was this man who would destroy a wild outcrop lush with mosses so that he might decorate his garden with the illusion of antiquity? Who was this man who bought time and who bought me? The Owner. What was this power of facelessness, so that no one uttered his name?

I am trying to understand what it means to own a thing, especially a wild and living being. To have exclusive rights to its fate? To dispose of it at will? To deny others its use? Ownership seems a uniquely human behavior, a social contract validating the desire for purposeless possession and control.

To destroy a wild thing for pride seems a potent act of domination. Wildness cannot be collected and still remain wild. Its nature is lost the moment it is separated from its origins. By the very act of owning, the thing becomes an object, no longer itself.

Blowing up a cliff to steal the mosses is a crime, but it's not against the law, because he "owns" those rocks. It would be easy to call this abduction an act of vandalism. And yet, this is also a man who imports a team of experts for the gentle wrapping of mossy rocks. The Owner is a man who loves mosses. And the exercise of power. I have no doubts of his sincerity in wishing to protect them from harm, once they conformed to his landscape design. But I think you cannot own a thing and love it at the same time. Owning diminishes the innate sovereignty of a thing, enriching the possessor and reducing the possessed. If he truly loved mosses more than control, he would have left them alone and walked each day to see them. Barbara Kingsolver writes, " It's going to take the most selfless kind of love to do right by what we cherish and give it the protection to flourish outside our possessive embrace".

When the Owner looks at his garden, I wonder what he sees. Perhaps not beings at all, only works of art as lifeless as the silenced drum in his gallery. I suspect that the true identities of mosses are invisible to him and yet he wanted authenticity more than anything. He was willing to pay huge sums to have authentic moss communities at his doorstep where his guests might praise his vision. But in possessing them, their authenticity is lost. Mosses have not chosen to be his companion, they have been bound.

I was dropped back at the staging area, with the coolness reserved for a team member who won't play the game. I was wearily walking to my car when I saw Matt getting into his truck. He was friendly enough and said that he'd been reassigned to another project. Mosses were no longer his responsibility, he said, with bright conviction in his face. But, knowing my interest, he wanted to show me just one more thing. He was on his way home, no longer on the Owner's time. So we got into his battered pickup and he clicked off his radio and its constant demands. We talked of his baby daughter and of azaleas but not of the terrace garden. He drove me through the woods to the far edge of the property along the security road. The boundary was marked by a four-strand electric fence angled outward to foil deer and other intruders. The entire perimeter beneath the fence had been treated with Roundup, killing all the vegetation. All ferns, wildflowers, shrubs, and trees had been eliminated in a ten-foot swath. All were dead except the mosses. Immune to the chemicals, the mosses had taken over, colonies coalescing into a crazy quilt in a thousand shades of green. Here was the real moss garden of the Owner, a mile from his house living under an electric fence in an herbicide rain.

The Forest Gives Thanks to the Mosses

From the windswept silence of Marys Peak, you can see the struggle unfolding. The land that stretches to the ocean, sparkling seventy miles distant, is broken into fragments. Patches of red earth, smooth blue green slopes, polygons of bright yellow green and amorphous dark green ribbons sit uneasily side by side. The Oregon Coast Range is a patchwork of clear-cuts and the second and third generations of Douglas-fir coming back with a vengeance. The landscape mosaic also includes a few scattered remnants of the original forest, the old growth that once stretched from the Willamette Valley to the sea. The landscape spread out before me looks more like ragged scraps than a patterned quilt. It looks like indecision as to what we want our forests to be.

The conifer forests of the Northwest are renowned for their abundance of moisture. The temperate rain forests of western Oregon receive as much as 120 inches of rain a year. The mild rainy winters let trees grow all year round and, with them, their mosses. Every surface in a temperate rain forest is covered with moss. Stumps and logs, the entire forest floor is greened with wildly tangled turfs of *Rhytidiadelphus* and translucent clumps of *Plagiomnium.* The tree trunks are feathered with plumes of *Dendroalsia* like the breast of a great green parrot. Vine maple shrubs arch over under the weight of curtains of *Neckera* and *Isothecium* two feet long. I can't help it, my heart beats faster when I come into these woods. Perhaps there is some intoxicating element in moss-breathed air, transmuted in its passage through glistening leaves.

Indigenous people of these forests, and all over the world, offer traditional prayers of thanksgiving which acknowledge the roles of fish and trees, sun and rain, in the well-being of the world. Each being with whom our lives are intertwined is named and thanked. When I say my morning thanks, I listen a moment for a reply. I've often wondered if the land any longer has reason to return gratitude toward humans. If forests

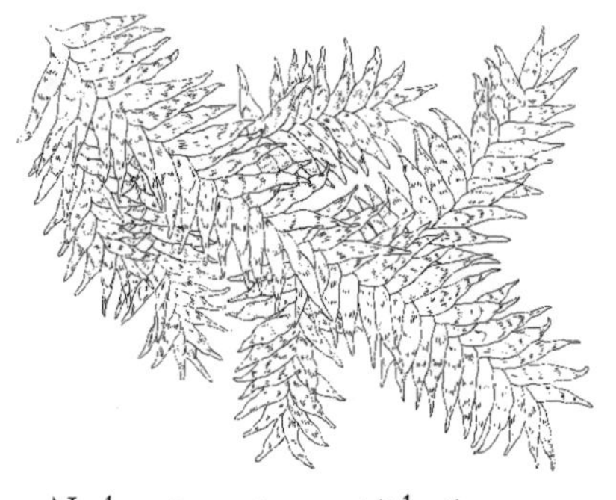

Neckera pennata, an epiphytic moss

did offer prayers, I suspect they would send thanks to the mosses.

The beauty of mosses in these forests is much more than visual. They are integral to the function of the forest. Mosses not only flourish in the humidity of a temperate rain forest, they play a vital role in creating it. When rainfall meets a forest canopy, its potential routes to the ground below are many. Very little precipitation actually falls directly to the forest floor. I've stood in a forest during a downpour and been as dry as if I had been holding an umbrella. The raindrops are intercepted by the leaves, where they slide off toward the twigs. At a junction, two drips meet and then two more, forming tiny rivulets at the confluence of branches. Like tributaries of an arboreal river, all flow toward the stream running down the trunk of the tree. Foresters call this water coursing down the tree "stemflow." "Throughfall" is the name for water which drips from branches and leaves.

I like to pull up my hood and stand close to a tree trunk in a rainstorm and watch the progress of the flood. The first droplets sink into the bark like rain into thirsty soil, the corky layers absorbing moisture. Then the gullies of the bark are filled to overflowing, until the water leaves their banks and sheets over the entire surface. Miniature Niagaras form over ledges in the bark, sweeping bits of lichen and helpless mites in the torrent. Passing over twig and branch it picks up sediment along the way. Dust, insect frass, microscopic debris, all are swept along, dissolving in the water so that stemflow is far richer in nutrients than the pure rainwater from which it began. In effect, the rain washes the trees and carries the bathwater straight to the waiting roots. This recycling of nutrients from the rinsed bark to the soil keeps the valuable nutrients in the tree and prevents their loss from the forest floor. The soil gives thanks to the mosses.

Like pillowy sandbags set in the way of a river, the clumps of moss also slow the passage of rain down the trunk of a tree. As water flows over mosses, much of it is absorbed into the tiny capillary spaces of the clump. Water is held in tapered leaf tips, funnelled into tiny drain pipes to the concave basins at the base of every leaf. Even the dead portions of the colony, the old leaves and tangled rhizoids, can trap moisture. The amount of rain held in Oregon mosses has never been measured, but in a mossy cloud forest in Costa Rica the mosses absorbed 50,000 liters of water per hectare of forest in a single rainfall. It's easy to see how flooding follows quickly on the heels of deforestation. Long after the rain has gone the mossy tree trunks remain saturated and slowly release last week's rain. When a shaft of light comes through the canopy and focuses on a clump of moss, you can watch the steam rise. Clouds give thanks to the mosses.

Mist rolls in each evening from the sea. High in the canopy, the mosses are poised to gather it like silver berries. The intricate surface of a moss colony becomes beaded with moisture as hair-like leaf points and delicate branches invite the condensation of fog droplets. In addition, the cell walls of mosses are rich in pectin, the same water-binding compound that thickens strawberries into jam. The pectin enables mosses to absorb water vapor directly from the atmosphere. Even without rainfall, the canopy mosses collect water and slowly drip it to the ground, keeping the soil moist for the growth of trees, which in turn sustain the mosses.

I like paper. I like it a lot—its weightless strength, its inviting blankness. I like how it waits, the clean white rectangle framed by the smooth oak of my desk. The oak grain ripples and catches the light like no petroleum byproduct ever could. I like the pine paneling in my cabin and the smell of woodsmoke on an autumn night. But despite my love affair with forest products, a log truck going by on the highway makes me sad, especially on a rainy day when clumps of moss still cling to the trunks, watered by the dirty spray of passing semis. Just days ago, when those logs were still trees, these same mosses were full of forest moisture and not the diesel wash thrown up by tires on I-5.

I can't help but poke at my own inconsistency, like a tongue probing at a loose tooth. I surround myself with real forest products, yet rail against the clear-cuts my desires create. In Oregon, clear-cuts are the "working forest," the blue-collar trees that yield my neat stack of paper and the roof of my house. I'm caught in the same conflict that we see on the fragmented landscape. I decided that I needed to confront my ignorance and go visit a clear-cut.

One bright Saturday morning my friend Jeff and I set out to drive to a Coast Range clear-cut. It isn't hard to find one. Uncut buffer strips are required between public thoroughfares and the cuts, protecting the public's view by federal mandate. Loggers complain about the trees left unharvested, but those thin, concealing walls of forest may serve the industry well, creating the roadside illusion of intact forest and stifling public objections. We turn down a new logging road, past the gate and warning signs. Here there is no concealing screen between you and the land. We nearly have to turn back. I rationalize my nausea as vertigo from the precipitous roads, the cold sweat as anxiety about oncoming log trucks. But I knew it was fear and the presence of violence around every turn. And grief, grief that rises up from the stumps and soaks into our skin.

It's a scene we all want to turn away from, but we had better look at the consequences of what we choose. Jeff and I lace up our hiking boots and start off across the slope. I'm looking hard for signs of remnant mosses, signs of incipient recovery. But all I see is a wasteland of stumps and tattered plants frizzled to a rusty brown in the intense glare of the sun. The luxuriant forest floor has been replaced by windrows of chips. The smell of damp earth is replaced by the fragrance of pitch oozing from cut stumps. It is hard to believe that the same amount of good clean rain falls on a clear-cut as on the adjacent patch of old-growth forest. The land is as dry as sawdust. All that water doesn't do much good without a forest to hold it. Streams draining clear-cut watersheds carry much more water than a stream running through a forest. And without the mossy forest to hold it, the water runs brown with soil, silting up salmon streams as it carries the land to the sea. Rivers give thanks to mosses.

This raw cut on the land will be replanted with Douglas-fir seedlings, a high-performance monoculture. But trees alone don't make a forest and many organisms have a tough time ever recolonizing the cut-over land. Mosses and lichens, so vital to forest function, disperse very slowly into a recovering forest. Forest scientists have made efforts at finding management practices which will encourage the return of forest biodiversity. Old logs must be left behind to provide habitat for mycorrhizae and salamanders, dead trees for woodpeckers. In a good-faith effort to speed the regrowth of epiphytes, forest policies now dictate that a few old trees must be left standing as refuges for the mosses that will colonize the new forest. It's a hopeful thought, that the coming monoculture of Doug fir will be seeded with mosses spreading from the few trees left behind. But first the remnant mosses, like desert islands in a sea of stumps, must survive the loss of the forest around them.

Far down the hillside, I see a lone survivor. A ribbon of striped flagging flutters in the hot wind. It's a marker for the logging crew that this individual was to be left standing to meet the obligation of law and to reseed the forest. I skid down the slope toward it, dodging a tangle of cut limbs. The rains have washed a gully down the hillside. I jump over and dust rises where I land. The survivor stands alone like the last person on earth. There's no joy in being spared the saw when everyone else is gone, riding down the highway to the mills at Roseburg.

I'd expected a pool of shade beneath the survivor, but its branches are so high up that any shadow is thrown way out among the stumps. Looking up into the canopy I can see that whoever marked this tree had chosen well. It is a prime example of the lush tree-top community that is the signature of an old forest. The trunk and branches are laden with the skeletons of mosses. The sun has bleached away their green and the brown mats are peeling away. Withered carcasses of fern rhizomes are exposed beneath the moss. The wind works at the loose edge of an *Antitrichia* mat, rustling. We stand there, wordless.

The poikilohydric nature of mosses allows many species to dry and then recover whenever water is available. But the species here, accustomed to the sweet and steady moisture of the forest, have been pushed beyond their limits of tolerance. Sunbaked and desiccated, they're unlikely to endure until the next forest returns. I'm encouraged that the makers of forest policy gave thought to the mosses and their presence in the future forest. But mosses are intertwined with the fabric of a forest and can't exist alone. If mosses are to take their place in the recovering forest, they must be granted a refuge that will sustain them. If given a voice, I think they would advocate for patches large enough to hold moisture, shady enough to nurture their entire community. What is good for mosses is also good for salamanders, waterbears, and wood thrushes.

ꕥ

There is a positive feedback loop created between mosses and humidity. The more mosses there are, the greater the humidity. More humidity leads inexorably to more mosses. The continual exhalation of mosses gives the temperate rain forest much of its essential character, from bird song to banana slugs. Without a saturated atmosphere, small creatures would dry out too quickly, due to their extravagantly high surface area to volume ratio. As the air dries, so do they. So without the mosses, there would be fewer insects and stepwise up the food chain, a deficit of thrushes.

The insects find shelter in the moss mat, but only rarely are the moss shoots actually eaten. Birds and mammals likewise avoid consuming them, with the exception of some large sporophytes which are high in protein. The almost total lack of herbivory on mosses may be due to the high concentration of phenolic compounds in the leaves, or perhaps their low nutritional value makes it unprofitable to eat them. The toughness of moss cell walls also makes them rather indigestible. Animals that do ingest mosses often pass them out again nearly intact. The indigestible fiber of mosses has been reported from a surprising location—the anal plug of hibernating bears. Apparently, just before entering the winter den, bears may eat a large quantity of moss, which so binds up their digestive system that it blocks defecation through the long winter sleep.

A whole array of insects spend their larval phases inching their way through moss mats, unseen until the moment of metamorphosis.

Wiggling out of old skins, they venture out on new wings into the moss-moistened air, free. They feed, they mate, and days later deposit their eggs in a mossy cushion and fly off. To be eaten by a hermit thrush, whose eggs lie cradled in a nest lined with mosses.

Soft and pliable, mosses are woven into birds' nests of many species, from the velvety cup of a winter wren to the hanging basket of a vireo. They find their greatest use in the bottom of the nest, to cushion the fragile eggs and to provide an insulating layer. I once found a hummingbird nest where trailing mosses decorated the rim of the tiny nest like fluttering Tibetan prayer flags. Birds giving thanks for mosses. They aren't the only ones who rely on mosses for nesting material; flying squirrels, voles, chipmunks, and many others line their burrows with bryophytes. Even bears.

The marbled murrelet is a coastal bird which feeds on the wealth of marine life along the Pacific shore. For decades, its numbers have been dwindling and it is now listed as an endangered species. The cause of its decline was unknown. Other coastal birds nest along the shore where food is plentiful, forming rookeries on rocky cliffs and seamounts. But murrelets never joined them. Their nesting sites were thought to be hidden since none had ever been seen. In fact, murrelets nest at the tops of old trees, far from their coastal feeding grounds. Every day the birds fly as much as fifty miles inland, to the old-growth forests of the Coast Range. Their disappearance was due primarily to the disappearance of the old growth. Researchers found that most murrelet eggs were laid in a nest made of *Antitrichia curtipendula*, a luxuriant golden green moss endemic to the Pacific Northwest. This pair, moss and murrelet, are both reliant on the old growth.

It seems as if the entire forest is stitched together with threads of moss. Sometimes as a subtle background weave and sometimes with a striking ribbon of color, a brilliant fern green. The ferns which decorate the trunks and branches of the old-growth trees are never rooted in bare bark, always in moss. Ferns give thanks for mosses. Licorice fern runs rhizomes beneath the moss, anchored in the accretion of organic soil.

Towering trees and tiny mosses have an enduring relationship that starts at birth. Moss mats often serve as nurseries for infant trees. A pine seed falling to the bare ground might find itself pummeled by heavy raindrops or carried off by a scavenging ant. The emerging rootlet may dry in the sun. But a seed falling on a bed of moss finds itself safely nestled among leafy shoots which can hold water longer than the bare soil and give it a head start on life. The interaction between seed and moss is not universally positive; tree seedlings may be inhibited if the seed is small and the moss is large. But often mosses facilitate the establishment of trees. Mossy logs are often referred to as "nurse logs." The remnants of that nurture can be seen in the straight lines of hemlocks sometimes found in the forest, a legacy of seedlings who shared a beginning on a moist log. Trees give thanks for mosses.

Moisture begets moss and moss begets slugs. The banana slug must be the unofficial mascot of the Pacific Northwest rain forests, gliding over mossy logs and surprising hikers with six inches of dappled yellow mollusk stretched across a trail. The slugs feed on the many inhabitants of a moss turf, and even on the moss itself. A biologist friend of mine, interested in all things small, once scooped up some slug feces while waiting for a bus and brought them home for a look under the microscope. Sure enough, they were full of tiny moss fragments, and he happily phoned me to report the good news. Slugs eat mosses and disperse them in return. Biologists may make unsuitable dinner conversation, but we are seldom bored.

Banana slugs are most abundant in the morning, when their slime trails still glisten on the logs. They seem to disappear by the time the dew has dried. But where do they go? I discovered their hideaway one afternoon when I was looking at the flora of decaying logs. Peeling away a layer of *Eurhynchium* from a massive log, I exposed what seemed to be a whole dormitory of banana slugs. Lying in individual rooms of spongy wood, each was nestled between the cool moist wood and the blanket of moss. I hastily covered them up, before the sun could catch them sleeping. Slugs give thanks to the mosses.

The logs of the forest floor shelter more than slugs and bugs, playing an integral role in the nutrient cycle of the ecosystem. The fungi responsible for decay reside there and their survival is highly dependent upon constant moisture in the log. The coating of mosses insulates the log from drying, providing an environment where the fungal mycelium can flourish. The thread-like mycelium is the hidden part of the fungus, the working equipment of decomposition. A wide variety of fungi are found only on deep mats of moss. The beautiful mushrooms are but the tip of the iceberg, the showy reproductive phase which sprouts up from logs like a tiny flower garden. Fungi give thanks to the mosses.

A specialized class of fungi, essential to forest function, also resides beneath the moss carpet of the soil. On the surface, scraggly turfs of *Rhytidiadelphus* and mops of *Leucolepis* cover the forest floor. Beneath them in the humus live the mycorrhizae, a group of fungi which live symbiotically with the roots of trees. The term literally means fungus (myco-) root (-rhizae). The trees host these fungi, feeding them the sugars of photosynthesis. In return, the fungi extend their filamentous mycelium out into the soil to scavenge nutrients for the tree. The vigor of many trees is completely dependent on this congenial relationship. It has recently been discovered that the density of mycorrhizae is significantly higher under a layer of mosses. Bare soil is far less hospitable to this partnership. The association of moss and mycorrhizae may be due to the even moisture and nutrient reservoir beneath the moss carpet.

Studying the interactions that happen belowground, among microscopic beings, is notoriously difficult, but a group of researchers has untangled an intricate three-way connection. Tracing the flow of phosphorus through a forest, they followed its footprints in a convoluted path that started with the rain. Throughfall washed phosphorus from the spruce needles onto the mosses below, where it was stored until mycorrhizal fungi insinuated their filaments into the moss turf. Their thread-like hyphae and extracellular enzymes absorbed phosphorus from the dead tissue of mosses. These very same fungi with hyphae in the moss also had hyphae in the roots of the spruce, forming a bridge between moss and tree. This web of reciprocity ensures that phosphorus is endlessly recycled, nothing wasted.

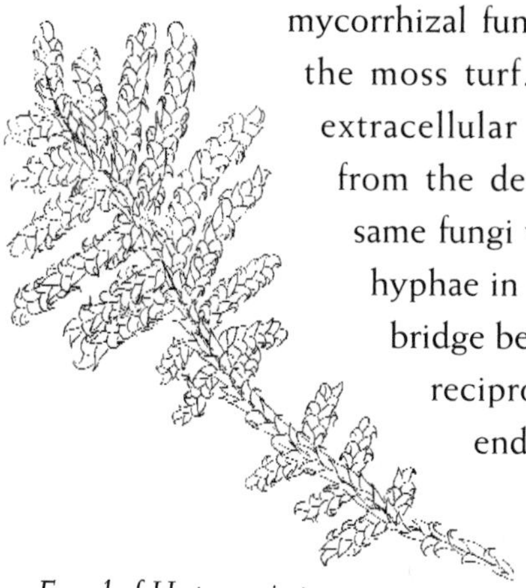

Frond of Hypnum imponens, common on mossy logs

The patterns of reciprocity by which mosses bind together a forest community offer us a vision of what could be. They

vision of what could be. They take only the little that they need and give back in abundance. Their presence supports the lives of rivers and clouds, trees, birds, algae, and salamanders, while ours puts them at risk. Human-designed systems are a far cry from this ongoing creation of ecosystem health, taking without giving back. Clear-cuts may meet the short-term desires of one species, but at the sacrifice of the equally legitimate needs of mosses and murrelets, salmon and spruce. I hold tight to the vision that someday soon we will find the courage of self-restraint, the humility to live like mosses. On that day, when we rise to give thanks to the forest, we may hear the echo in return, the forest giving thanks to the people.

The Bystander

Digging my boots into the hill, I gather my strength and lunge for the next handhold, a clump of stems above me. A thorn plunges deep in my thumb, but I can't let go. This is my only anchor. The bright blood welling up around it draws my attention to something other than the ache in my legs and the sound of my heart in my ears. Why in the world would they come all the way up here? The tangle of salmonberry is so thick in places that I can't break through. I've had to crawl on my hands and knees to find a route under it. The thorns constantly grab at my hat and my pack and my skin. My clothes are so heavy with mud, every step is an effort. And now I've lost whatever remnant of a trail they left. I feel myself on that knife-edge between laughter and crying. Exhausted, I cast about for some excuse to give up the search and get out of here. Then, out of the corner of my eye I catch sight of red, tattered flagging tied to a branch upslope. That must be the way they came. I'll bet they marked a path for a quick getaway when they were done. My thumb tastes like mud and iron as I suck off the blood and push forward, shielding my face from the brambles with every lunge.

The higher I go I'm enveloped more and more in the mist that caps these Coast Range hilltops. The gray only adds to the chill and to the growing realization of how far I've come. And that nobody else knows exactly where I am. Not even me. The sound of an agitated pack of hunting dogs from the valley floor makes me realize I'm not alone. My presence is now known. I grimly hope that they don't decide to come investigate this trespasser. That's all I need. I have as much right to be here on public land as they do, but that would hardly matter. Those dogs probably came with them, and lay with their tongues hanging out as they watched.

At the lip of the hill, it suddenly flattens out into a stand of mist-covered maples. My heart slows down just a fraction and I try to wipe the sweat from my eyes with a muddy hand. The salmonberries thin out and I can see more than a few feet ahead of me again. I know instantly that this is the place. So these are the riches that drew them up that insufferable hill. They'd discovered the mother lode. Besides, it's remote enough that they'd never get caught. It's been a while since they were here and the place still shows the hand of violence.

Once they finally got here, I guess it must have been easy pickings. It's thick up here where the mist hangs on the hill all day. They must have filled the sacks they brought quicker than they expected, since the stand is only half stripped. They'd never guess there would be so much, and it's heavy to carry.

The woods across the stream seem to be untouched. The vine maples there are hung with sheets so thick the air itself looks green. There's not a single spot that isn't covered with mosses. I know what I'd see if I looked close. The amazing stuff that only these remote old stands still have—every one an old friend of mine. You don't see these big feathers of *Dendroalsia* much any more, or the clumps of *Antitrichia* so thick you could sink your hand into them. Shining ropes of *Neckera*. And so much more. I wince when I think they probably didn't even stop to look. At least art thieves know what they're taking.

The other side of the stand has been picked clean; like vultures, they left only the bare bones. I imagine them sticking their dirty hands deep into the mat and ripping it off in swaths the length of their arms. It gives me the shivers to think of that tearing, like a woman stripped naked before her attackers. Peeling back the moss from tree after tree, they cram them all into the burlap bags, light into dark. You've got to acknowledge that they're efficient predators. The exposed bark is utterly naked.

It bothers me that they sat here and had a self-satisfied cigarette after their work was done. They left the package stuffed into the hollow of a log. I imagine that they whistled to the dogs and headed back down the hill, dragging their hostages behind them. That must have been as bad as coming up the hill, with the salmonberry grabbing at the bags. I can't blame them for not coming back to finish the job. A pickup load's not bad for the day. There's a buyer down at the Pacific Pride station who's paying cash.

And now my work begins, to catalog the aftermath. I feel like the photographer helplessly documenting a disaster, passive and unable to change the outcome. We find these spots where the moss pickers have been and become scientific witnesses to the destruction. Every scalped branch will be measured and tagged and examined for signs of re-growth. I'm looking hard for hope that these naked branches will have started to green up again. They haven't. Maybe a trailing shoot and here and there a lone branch ventures out onto the hard dry bark. The recovery is almost nonexistent. You don't need a sophisticated analysis to see that, but I dutifully record my data. No one knows how long it will take for them to grow back. Maybe never. Most of these mats were as old as the trees themselves, and got started when they were just saplings.

The intact part of the stand, though, yields enough measurements to stuff my data book as full as the moss pickers' sacks. Every branch has at least a dozen species of moss, in a dozen different shades of green. *Eurhynchium, Claopodium, Homalothecium* ... each of them a work of art, a marriage of light and water to produce a carpet more intricate than anything on the planet. An antique tapestry ripped to shreds and stuffed in a bag. And in the bag are also untold billions of beings who made that moss their home, like birds nesting in a forest. Scarlet Orabatid mites, bouncing springtails, whirling rotifers, reclusive waterbears, and their children: shall I say all their names in a requiem mass?

All this destruction—for what? If we followed the pickup truck down to the city, we could watch them heave their prizes up on the scale at the loading dock and walk away with pockets a little heavier, but not much. At the warehouse, the sacks are dumped out and their contents cleaned and dried. This premium product, "Oregon Green Forest Moss," is marketed all over the world. The marketers trade on the Oregon name to invoke an image of lush forests. Depending on the species and the quality, the mosses are sorted for different products. Low-grade material goes to line flower baskets sold to florists, or to dress up synthetic greenery with what the catalog calls "A Lifelike Look." The most robust and beautiful are saved for a special treatment—the creation of "Designer Moss Sheets." Feathery fronds are glued to a fabric backing and then sprayed with flame retardant so that they will meet fire codes in public places, like beneath the motorcycles at an auto show, and in the most elegant hotel lobbies. The finishing touch is a patented process that applies the trademarked "Moss Life" dye to create a vivid green. This moss "fabric" is rolled up in bolts, ready for sale. Sheets are sold by the yard and the website advertises that it can be used "wherever Mother Nature's touch is wanted."

I saw them in the main concourse at the Portland airport, filling in the spaces under the plastic trees. I breathed their names when I saw them—*Antitrichia, Rhytidiadelphus, Metaneckera*—but they turned their eyes away.

ꟹ

The rainy forests of the Pacific Northwest create ideal conditions for moss growth. The branches of shrubs and trees are often draped with thick mats of epiphytes, containing many species of mosses, liverworts, and lichens that play important roles in nutrient cycling, food webs, biodiversity, and habitat for invertebrates. The weight of living mosses is estimated at between ten and two hundred kilograms per hectare. In some forests, the weight of mosses may exceed the weight of tree leaves.

Since 1990, this luxuriant moss growth has come under attack from commercial moss harvesters, who strip branches completely bare and sell the moss to the horticultural industry. Legal moss harvest in the Coast Range of Oregon has been estimated to exceed 230,000 kilograms per year. The Forest Service regulates moss harvest on National Forests by a system of permits, but enforcement is minimal. Illegal harvest is thought to be as much as thirty times higher than the legal quota. Additional quantities are taken from other public and private forests.

Bryologists have been studying some experimentally harvested plots in order to estimate how quickly the mosses will regrow. Our preliminary studies suggest that recovery may require decades. Four years after harvest, the branches of the vine maple are smooth and bare, with scarcely a trace of moss returning. The mosses left at the torn edges of the stripped branch still cling, but creep out onto the bare area at far less than a snail's pace—a few centimeters in four years. We've discovered that the smooth mature bark is simply too smooth and slippery for mosses to get a foothold.

Kent Davis and I began looking at how mosses get started naturally as epiphytes. Surely they must be able to colonize bare bark—otherwise how could these thick moss carpets ever develop? What we found surprised us. Mosses on young trees don't colonize the bare bark at all. When we looked at tiny twigs and young branches, the bark was bare. But on virtually every leaf scar, bud scar, and lenticels there was a tiny tuft of moss. If you look closely at a twig, you'll see that much of it is covered by bark, but it also is textured by its brief history. It is marked by a raised stub where last year's leaf was held. This so-called leaf scar is minutely corky, with just enough texture to capture a spore or two. Twigs also bear a collection of closely spaced ridges which show where the bud used to be. This roughening also seems to provide a foothold for mosses. A young branch starts to collect its moss flora tuft by tiny tuft, leaf scar by leaf scar. We observed that the size of the tufts increased with the age of the branch. As the tree grew older, different mosses would colonize—not on the bare bark, but on top of the initial mosses. The thick moss mats of mature trees got their start on twigs. We found that colonization is much easier on rough young twigs and nearly impossible on old branches. Once the stem has aged and leaf scars become few and far between, the opportunities for attracting mosses have dwindled. We can infer, then, that the mats of mosses which weigh down the branches are probably almost as old as the trees themselves.

Moss harvesters are in a sense removing "old-growth" mosses, which cannot replace themselves nearly as quickly as they are removed. This is, by definition, unsustainable harvest. Their loss will have consequences we cannot foresee. When the mosses are taken, their web of interactions goes along with them. Birds, rivers, and salamanders will miss them.

This spring I was buying some perennials at my local nursery in upstate New York, a continent away from the mossy forests of Oregon. The displays in the garden shop were enticing as always, sundials and beautiful crockery. As we browsed among them my daughter caught my arm and said ominously "Look." Lined up against the wall was a menagerie of topiary beasts: life-sized reindeer, green teddy bears, and graceful swans. Every one a wire skeleton stuffed with the carcasses of Oregon moss. The time to be a bystander has passed.

Straw into Gold

It disappeared the year I brought curtains. I knew it was a mistake, yet having made them, owned them, I felt strangely bound to let them hang, although they tangle in the wind and plaster themselves wet against the screen in a thunderstorm. Such is the tyranny of possessions. The window swings inward, a big eight-pane square of wavy glass with the glazing weatherbeaten and falling out in chunks. I almost never close it, night or day. Through that window comes the incessant sound of the lake, and the smell of white pines, resinous in the sun. Why would a person hang curtains in the wilderness? To shut out starlight on a black, black night? To prevent the looking in by a thousand pinpoint stars?

Every spring I lock the door on my house full of things, the feathered nest of books and music, soft lights, comfy chairs, and—I blush to admit it—the three computers and a dishwasher. I drive away from the carefully tended gardens, with delphiniums just coming in to bloom, taking with me as little as possible. As I drive northward, on my annual migration from the rolling farmlands of upstate New York to the unbroken forest of the Adirondacks, the comfortable life of the professor's house slips farther and farther away.

The Biological Station is an outpost on the far eastern shore of Cranberry Lake. It is accessible only by a seven-mile boat trip across the open lake. In early June, the crossing can be a rough one. After all, the water was ice only six short weeks ago. The rain and the waves join forces and sheet from my raincoat sleeves. I turn to look at the girls, huddled in the stern with their heads tucked in their ponchos like red and blue turtles. The wind nearly blows my glasses off my face, and I'm blinded in the rain, trying to keep the boat abreast of the waves. One slap of the bow against a roller and we are drenched. The icy water finds its way into the small unzipped gap at my throat and trickles down between my breasts. All we have is in this boat. All we need lies on the shore ahead.

We arrive at the dock as the skies deepen to darkness and walk up through the dripping woods to the unlit cabin, just visible in the steely gray reflection from the lake. We peel off our wet clothes in the dark and I fumble to find the coffee can of matches The kids stand close behind me, in blankets, as I kneel at the fireplace. Their wet socks leave footprints on the floor. The first sulfur flare of the match seems to light up the whole room, first blue and then golden as it catches the birch bark. For me, the incense of yellow birch bark has always been the scent of safety. I breathe a sigh of relief and the tension rolls off my tired shoulders like the rain from the roof. On this far shore, on this rainy night with the firelight dancing on the bare walls, I am more content than in my warm house full of lovely things. Here is every single thing I need. And it is precious little: rain on the outside, fire on the inside. And soup. And the rest is luxury. Especially curtains.

Every summer I brought less and less. When they were little, the girls could each bring just one toy and a rainy-day box full of crayons, paper, and such. But they usually went back home unused. A whole summer was not nearly long enough for all the rocks there were to climb and forts to build. The crayons languished while little villages of pebbles and pine cones spread out under the pines. They tied blue jay feathers in their pigtails and ate up the summer in heaping spoonfuls like homemade peach ice cream. After supper I'd set aside my moss work and we'd go scrambling along the lake edge. Late in the long day, the low hanging sun across the lake would bathe our shore in light as thick and gold as honey. We'd clamber over rocks and get our feet wet, dodging the waves. The girls would intently examine bits of driftwood and pearly mussel shells, their faces luminous in the sunset, shining gold. This is when we saw it, that most improbable of beings.

The fires that burned here at the turn of the twentieth century gave us a lakeside border of paper birches, brilliant white and rooted in the glacial sands. The last glacier gave us a lakeshore of granite boulders. The jumble of rocks make solitary spots for watching the sunset and create a strong barrier to the action of wind and waves. But there are breaches among the rocks where the waves have ridden in on a storm, undermining the sandy shore and excavating small caves. We poke our heads into the caves, brushing away the sticky spider webs that stretch over the entrances. The caves are small enough for kids to crouch inside, but grownups are barred. We can only look. I lie on the lake-washed cobbles, my head in the cave, looking up into the dimness. It smells cool and moist like the dirt floor of an old cellar. The sounds of the waves are muffled inside, my daughters' excited breathing seems loud in the dark quiet.

The roof of the cave is a dark dome, sand laced together with a network of birch roots. The back of the cave disappears upward into shadow. What light there is moves eerily, reflections from the water outside wavering up and down the cave walls. And then, out of the corner of my eye, something glitters. Something green. Something fleeting, like the eye of a bobcat in the firelight.

I stretch out my fingertip toward the green shimmer and pull it away when all I feel is dampness, like a film of cold sweat. I half expect my finger to glow like the time I accidentally squashed a firefly in the screwtop of a Mason jar on a summer night. But there's nothing. The soil surface itself seems to give off light. As I turn my head, the light comes and goes, glittering like the iridescence on a hummingbird's throat, one moment sparkling, the next moment black.

Schistostega pennata, the Goblins' Gold, is unlike any other moss. It is a paragon of minimalism, simple in means, rich in ends. So simple you might not recognize it as a moss at all. The more typical mosses on the bank outside spread themselves out to meet the sun. Such robust leaves and shoots, though tiny, require a substantial amount of solar energy to build and maintain. They are costly in the solar currency. Some mosses need full sun to survive, others favor the diffuse light of clouds, while *Schistostega* lives on the clouds' silver lining alone. Inside the shoreline caves light is reduced to mere watery reflections from the lake surface. It is only one-tenth of one percent of the light outside.

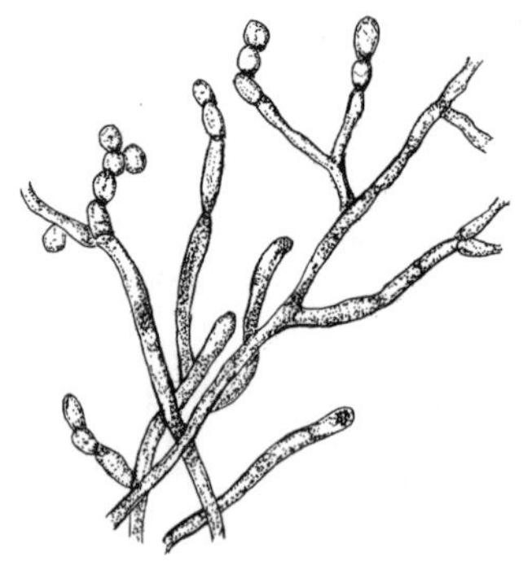

Filamentous protonema of Schistostega pennata

Sunlight in caves is much too scarce for *Schistostega* to afford much in the way of architecture. Leaves are a bit of a luxury in so spare an environment. So, in place of leaves and shoots, Goblins' Gold is reduced to a fragile mat of translucent green filaments, the protonema. The shimmering presence of *Schistostega* is created entirely by the weft of nearly invisible threads crisscrossing the surface of the moist soil. It glows in the dark, or rather it glitters in the half light of places which scarcely feel the sun.

Each filament is a strand of individual cells strung together like beads shimmering on a string. The walls of each cell are angled, forming interior facets like a cut diamond. It is these facets which cause *Schistostega* to sparkle like the tiny lights of a far-away city. These beautifully angled walls capture traces of light and focus it inward, where a single large chloroplast awaits the gathering beam of light. Packed with chlorophyll and membranes of exquisite complexity, the chloroplast converts the light energy into a stream of flowing electrons. This is the electricity of photosynthesis, turning sun into sugar, spinning straw into gold.

Here on the shadowy edge of where green life seems barely possible, *Schistostega* has all it needs. Rain on the outside, fire on the inside. I feel a kinship with this being whose cold light is so different from my own. It asks very little from the world and yet glitters in response. I have been blessed by the companionship of good teachers and I count *Schistostega* among them.

My small daughter blows at the roots dangling in front of her face. She looks like a goblin herself, crouched in the darkness, guarding the gold. Outside, the sun drops lower. A wide ribbon of orange light unrolls over the lake toward us. The sun is just a degree or two above the horizon now, its rim barely touching the hills on the opposite shore, sinking. The time is almost here. We're both holding our breath as the light starts to climb the walls of the cave. At last the sun drops low enough to reach the opening in the bank. Suddenly the sun pierces the darkness like a shaft of light through a slit in an Incan temple on the dawn of the summer solstice. Timing is everything. Just for a moment, in the pause before the earth rotates us again into night, the cave is flooded with light. The near-nothingness of *Schistostega* erupts in a shower of sparkles, like green glitter spilled on the rug at Christmas. Each cell of the protonema refracts the light, transforming it to the sugar that will sustain it through the coming darkness. And then, within minutes, it's gone. All its needs are met in an ephemeral moment at the end of the day when the sun aligns with the mouth of the cave. We climb back up to the top of the bank and walk back to the cabin as the sunset fades to dark.

On those most opportune of summer evenings light is plentiful. *Schistostega* responds by making more of itself to intercept the summer light. All along the protonema, tiny buds have been poised to take advantage of this transitory abundance. The buds expand to form ranks of upright shoots, scattered over the protonema. Each shoot is shaped like a feather, flat and delicate. The soft blue green fronds stand up like a glade of translucent ferns, tracking the path of the sun. It is so little. And yet it is enough.

The knowledge of this patch of moss was a gift to me and I share it judiciously. My old professor showed it to me before he retired, after he knew my fate was sealed as a bryologist. I wouldn't show it to just anyone. I'm afraid I was rather haughty about this knowledge, doling it out only to those who'd proven themselves sufficiently appreciative to be worthy of the gift. It's not that I'm afraid that they might value it so highly they would take it. Rather, I'm afraid that they might not value it enough. So I hoarded that gold, protecting it, I thought, from disrespect from someone for whom its tiny glitter was not enough.

The patient gleaming of light eventually earns *Schistostega* enough energy to support a family. In the moisture condensed on the walls of the cave, the sperm swim blindly until reaching the receptive female, and a sporophyte is born. The tiny little capsule arises from the base of a filmy frond, and casts its spores upon the breeze. I suspect that the offspring don't usually escape the breathless confines of the cave, and yet there are colonies of *Schistostega* scattered all along the shore. Somehow they find their way to other pockets of this accidental habitat. It's a good thing, because caves don't live forever.

My girls got older and had better things to do than wander along the shoreline at sunset. And without them, I visited the caves less and less often. I got busy with other things, like hanging curtains. That was the year the glowing moss disappeared. One evening, walking alone, I saw that the bank where it lived had slumped, collapsing under its own

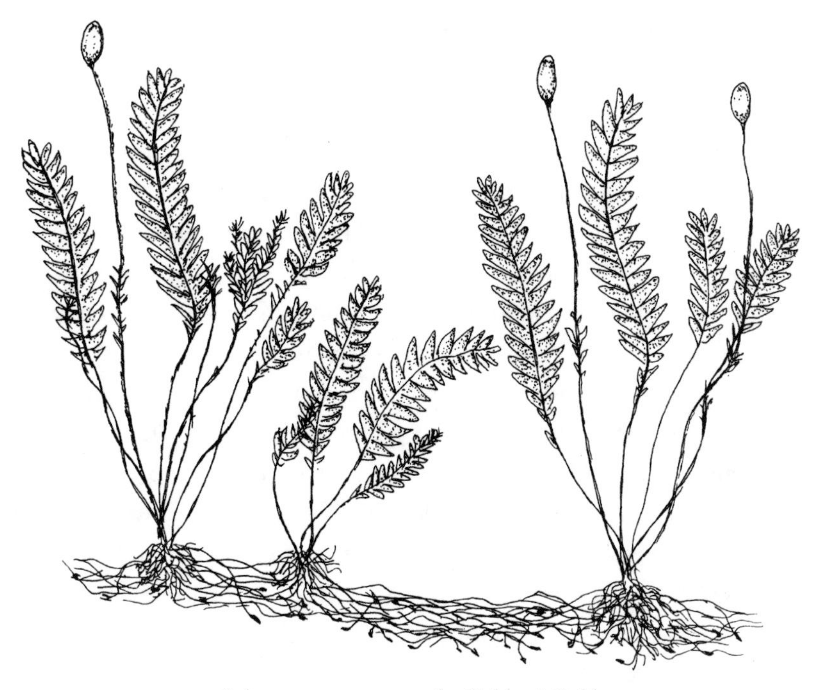

Schistostega pennata, the Goblins' Gold

weight and closing off the mouth of the cave. I suppose it was just the inevitable consequence of time and erosion. But I wonder.

An Onondaga elder once explained to me that plants come to us when they are needed. If we show them respect by using them and appreciating their gifts they will grow stronger. They will stay with us as long as they are respected. But if we forget about them, they will leave.

The curtains were a mistake. As if the sun and the stars and a moss that glitters were not enough to make it a home. Their superfluous flapping was a lapse in respect, a slap in the face to the light and the air waiting outside my window. Instead I invited in the small tyranny of things and let it make me forgetful. Forgetful that all I need is already here, rain on the outside, fire on the inside. *Schistostega* would not have made the same mistake. Too late, after the caves had collapsed, I threw the curtains into the woodstove and sent them up the chimney to the glittering stars.

Later that night, when the fire has died to embers and the moonlight is pouring in my window, I wonder about *Schistostega*. Can reflected moonlight set it to sparkling too? How many days a year can it rely on the sun to align with its window on the lake? Can it live on the opposite shore where it waits for the light of sunrise? Perhaps it's only here on our shore where the winds cut out caves and the sun can make a direct path between the rocks. The combination of circumstances which allows it to exist at all are so implausible that *Schistostega* is rendered much more precious than gold. Goblins' or otherwise. Not only does its presence depend on the coincidence of the cave's angle to the sun, but if the hills on the western shore were any higher the sun would set before reaching the cave. But for that small fact there would be no glitter. And only by virtue of the westerly winds steadily beating against the shore are there caves for *Schistostega* at all. Its life and ours exist only because of a myriad of synchronicities that bring us to this particular place at this particular moment. In return for such a gift, the only sane response is to glitter in reply.

Suggestions for Further Reading

BOOKS: Biology of Mosses

Bates, J. W., and A. M. Farmer, eds. 1992. *Bryophytes and Lichens in a Changing Environment.* Clarendon Press.

Bland, J. 1971. *Forests of Lilliput.*Prentice Hall.

Grout, A. J. *Mosses with Hand-lens and Microscope*

Malcolm, B., and N. Malcolm. 2000. *Mosses and Other Bryophytes: An Illustrated Glossary* Micro-optics Press.

Schenk, G. 1999. *Moss Gardening.*Timber Press.

Schofield, W. B. 2001. *Introduction to Bryology.*The Blackburn Press.

Shaw, A. J., and B. Goffinet. 2000. *Bryophyte Biology.* Cambridge University Press.

Smith, A. J. E., ed. 1982. *Bryophyte Ecology.*Chapman and Hall.

BOOKS: Identification of Mosses

Conard, H. S. 1979. *How to Know the Mosses and Liverworts.* McGraw-Hill.

Crum, H. A. 1973. *Mosses of the Great Lakes Forest.* University of Michigan Herbarium.

Crum, H. A., and L. E. Anderson. 1981. *Mosses of Eastern North America.* Columbia University Press.

Lawton, Elva. 1971. *Moss Flora of the Pacific Northwest.* The Hattori Botanical Laboratory.

McQueen, C. B. 1990. *Field Guide to the Peat Mosses of Boreal North America.* University Press of New England.

Schofield, W. B. 1992. *Some Common Mosses of British Columbia.* Royal British Columbia Museum.

Vitt, D. H., et al. *Mosses, Lichens and Ferns of Northwest North America.*Lone Pine Publishing.

OTHER RESOURCES

Alexander, S. J., and R. McLain. 2001. "An overview of non-timber forest products in the United States today." Pp. 59-66 in Emery, M. R., and McLain, R. J. (eds.), *Non-timber Forest Products.* The Haworth Press.

Binckley, D., and R. L. Graham 1981. "Biomass, production and nutrient cycling of mosses in an old-growth Douglas-fir forest." *Ecology* 62:387-89.

Cajete, G. 1994 *Look to the Mountain: An Ecology of Indigenous Education.* Kivaki Press

Clymo, R. S., and P. M. Hayward. 1982 The ecology of Sphagnum. Pp. 229-90 in Smith, A. J. E. (ed.), *Bryophyte Ecology*. Chapman and Hall.

Cobb R. C., Nadkarni, N. M., Ramsey, G. A., and Svobada A. J. 2001. "Recolonization of bigleaf maple branches by epiphytic bryophytes following experimental disturbance." *Canadian Journal of Botany* 79:1-8.

DeLach, A. B., and R. W. Kimmerer 2002. "Bryophyte facilitation of vegetation establishment on iron mine tailings in the Adirondack Mountains." *The Bryologist* 105:249-55.

Dickson, J. H. 1997. "The moss from the Iceman's colon." *Journal of Bryology* 19:449-51.

Gerson, Uri. 1982. "Bryophytes and invertebrates." Pp. 291-332 in Smith, A. J. E. (ed.), *Bryophyte Ecology*. Chapman and Hall.

Glime, J. M. 2001. "The role of bryophytes in temperate forest ecosystems." *Hikobia* 13: 267-89

Glime, J. M., and R. E. Keen. 1984. "The importance of bryophytes in a man-centered world." *Journal of the Hattori Botanical Laboratory* 55:133-46.

Gunther, Erna. 1973. *Ethnobotany of Western Washington: The Knowledge and Use of Indigenous Plants by Native Americans.* University of Washington Press.

Kimmerer, R. W. 1991a. "Reproductive ecology of *Tetraphis pellucida*: differential fitness of sexual and asexual propagules." *The Bryologist* 94(3):284-88.

Kimmerer, R. W. 1991b. "Reproductive ecology of *Tetraphis pellucida*: population density and reproductive mode." *The Bryologist* 94(3):255-60.

Kimmerer, R. W. 1993. "Disturbance and dominance in *Tetraphis pellucida*: a model of disturbance frequency and reproductive mode." *The Bryologist* 96(1)73-79.

Kimmerer, R. W. 1994. "Ecological consequences of sexual vs. asexual reproduction in *Dicranum flagellare*." *The Bryologist* 97:20-25.

Kimmerer, R. W., and T. F. H. Allen. 1982. "The role of disturbance in the pattern of riparian bryophyte community. *American Midland Naturalist* 107:37-42.

Kimmerer, R. W., and M. J. L. Driscoll. 2001. "Moss species richness on insular boulder habitats: the effect of area, isolation and microsite diversity."*The Bryologist* 103(4):748-56.

Kimmerer, R. W., and C. C. Young. 1995. "The role of slugs in dispersal of the asexual propagules of *Dicranum flagellare*." *The Bryologist* 98:149-53.

Kimmerer, R. W., and C. C. Young. 1996. "Effect of gap size and regeneration niche on species coexistence in bryophyte communities." *Bulletin of the Torrey Botanical Club* 123:16-24.

Larson, D. W., and J.T. Lundholm. 2002. "The puzzling implication of the urban cliff hypothesis for restoration ecology." *Society for Ecological Restoration News* 15: 1.

Marino, P. C. 1988 "Coexistence on divided habitats: Mosses in the family Splachnaceae." *Annals Zoologici Fennici* 25:89-98.

Marles, R. J., C. Clavelle, L. Monteleone, N. Tays, and D. Burns. 2000. *Aboriginal Plant Use in Canada's Northwest Boreal Forest.* UBC Press.

O'Neill, K. P. 2000. "Role of bryophyte dominated ecosystems in the global carbon budget." Pp 344-68 in Shaw, A. J., and B. Goffinet (eds.), *Bryophyte Biology.* Cambridge University Press.

Peck, J. E. 1997. "Commercial moss harvest in northwestern Oregon:describing the epiphytic communities." *Northwest Science* 71:186-95.

Peck, J. E., and B. McCune 1998. "Commercial moss harvest in northwestern Oregon: biomass and accumulation of epiphytes." *Biological Conservation* 86: 209-305.

Peschel, K., and L. A.Middleman. *Puhpohwee for the People: A Narrative Account of Some Uses of Fungi among the Anishinaabeg*. Educational Studies Press.

Rao, D. N. 1982. Responses of bryophytes to air pollution. Pp 445-72 in Smith, A. J. E. (ed.), *Bryophyte Ecology.* Chapman and Hall.

Vitt, D. H. 2000. "Peatlands: ecosystems dominated by bryophytes." Pp 312-43 in Shaw, A. J., and B. Goffinet eds. *Bryophyte Biology*. Cambridge University Press.

Vitt, D. H., and N. G. Slack. 1984. "Niche diversification of Sphagnum in relation to environmental factors in northern Minnesota peatlands." *Canadian Journal of Botany* 62:1409-30.

Index